Climate Change: Impacts and Management Strategies

Climate Change: Impacts and Management Strategies

Edited by Vivian Moritz

SYRAWOOD
PUBLISHING HOUSE

New York

Published by Syrawood Publishing House,
750 Third Avenue, 9th Floor,
New York, NY 10017, USA
www.syrawoodpublishinghouse.com

Climate Change: Impacts and Management Strategies
Edited by Vivian Moritz

International Standard Book Number: 978-1-64740-127-6 (Hardback)

Cataloging-in-Publication Data

Climate change : impacts and management strategies / edited by Vivian Moritz.
    p. cm.
Includes bibliographical references and index.
ISBN 978-1-64740-127-6
1. Climatic changes. 2. Climatic changes--Effect of human beings on.
3. Climatic changes--Management. 4. Climate change mitigation.
5. Environmental impact analysis. I. Moritz, Vivian.
QC903 .C55 2022
363.738 74--dc23

TABLE OF CONTENTS

PREFACE

Climate change refers to the changes in the statistical properties of climate system. The term is used to refer to anthropogenic climate changes, also known as global warming. Climate change includes both global warming as well as its effects such as precipitation that differs from one region to another. Global warming refers to the increase in average temperature of the earth's surface and the increased levels of greenhouse gases such as carbon dioxide, methane and nitrous oxide. The dominant cause of climate change is human activities like excessive emission from fossil fuel combustion such as coal, petroleum, diesel, natural gas, etc., carbon dioxide released by cement manufacture and deforestation. Alleviation of climate change includes the reduction of greenhouse gas release, the use of renewable energy, nuclear energy, enhancement of carbon sinks through reforestation. This book unravels the recent studies on climate change, and its impacts. It elucidates new management techniques and their applications in a multidisciplinary manner. This book will provide comprehensive knowledge to the readers.

This book unites the global concepts and researches in an organized manner for a comprehensive understanding of the subject. It is a ripe text for all researchers, students, scientists or anyone else who is interested in acquiring a better knowledge of this dynamic field.

I extend my sincere thanks to the contributors for such eloquent research chapters. Finally, I thank my family for being a source of support and help.

Editor

Adoption of multiple climate-smart agricultural practices in the Gangetic plains of Bihar, India

Jeetendra Prakash Aryal
*International Maize and Wheat Improvement Center (CIMMYT),
El Batan, Mexico*

M.L. Jat and Tek B. Sapkota
International Maize and Wheat Improvement Center (CIMMYT), New Delhi, India

Arun Khatri-Chhetri
*CGIAR Research Program on Climate Change Agriculture and Food Security
(CCAFS), Borlaug Institute for South Asia (BISA), New Delhi, India*

Menale Kassie
International Centre of Insect Physiology and Ecology (ICIPE), Nairobi, Kenya

Dil Bahadur Rahut
*Socioeconomics Program, International Maize and Wheat Improvement Center
(CIMMYT), El Batan, Mexico, and*

Sofina Maharjan
International Maize and Wheat Improvement Center (CIMMYT), New Delhi, India

Abstract

Purpose – The adoption of climate-smart agricultural practices (CSAPs) is important for sustaining Indian agriculture in the face of climate change. Despite considerable effort by both national and international agricultural organizations to promote CSAPs in India, adoption of these practices is low. This study aims to examine the elements that affect the likelihood and intensity of adoption of multiple CSAPs in Bihar, India.

Design/methodology/approach – The probability and intensity of adoption of CSAPs are analyzed using multivariate and ordered probit models, respectively.

Findings – The results show significant correlations between multiple CSAPs, indicating that their adoptions are interrelated, providing opportunities to exploit the complementarities. The results confirm that both the probability and intensity of adoption of CSAPs are affected by numerous factors, such as demographic characteristics, farm plot features, access to market, socio-economics, climate risks, access to extension services and training. Farmers who perceive high temperature as the major climate risk factor are more likely to adopt crop diversification and minimum tillage. Farmers are less likely to adopt site-specific nutrient management if faced with short winters; however, they are more likely to adopt minimum tillage in this case. Training on agricultural issues is found to have a positive impact on the likelihood and the intensity of CSAPs adoption.

Practical implications – The major policy recommendations coming from of our results are to strengthen local institutions (public extension services, etc.) and to provide more training on CSAPs.

Originality/value – By applying multivariate and ordered probit models, this paper provides some insights on the long-standing discussions on whether farmers adopt CSAPs in a piecemeal or in a composite way.

Keywords Climate change adaptation, Climate smart agricultural practices, Crop diversification, Minimum tillage, Site-specific nutrient management, Stress-tolerant seed varieties

1. Introduction

Agricultural production in India has increased considerably since the Green Revolution (GR) because of the adoption of high-yielding varieties, chemical fertilizer, pesticides, irrigation and mechanization. Sustaining the gains of the GR is becoming increasingly difficult because of negative environmental externalities, such as groundwater depletion, soil fertility degradation and chemical runoff (Pingali, 2012; Singh, 2000) resulting from resource-intensive agriculture. Increasing climatic variabilities exert further pressure on the sustainability of the existing production system. Recently, climate-smart agricultural practices (CSAPs) have been accepted as methods to address challenges because of climate change in Indian agriculture. National and international research organizations, donors and policymakers have been expending considerable resources to develop and promote these practices to increase agricultural productivity to feed the growing population and improve farmers' resilience to climate change. National Innovations in Climate Resilient Agriculture, National Mission for Sustainable Agriculture, Cereal System Initiatives in South Asia, Sustainable and Resilient Farming System Intensification and the CGIAR research program on Climate Change, Agriculture and Food Security (CCAFS) are some of these initiatives. Recent studies showed that most of the CSAPs have clear economic (Aryal *et al.*, 2015; Erenstein *et al.*, 2008; Khatri-Chhetri *et al.*, 2016; Sapkota *et al.*, 2015) and climate change adaptation benefits (Aryal *et al.*, 2016; Sapkota *et al.*, 2015). Despite the benefits of CSAPs and the considerable efforts by several organizations to promote them, the rate of their adoption by Indian farmers is still very low. Therefore, understanding the CSAPs adoption behavior of farm households is crucial to fine-tuning their design and promotion to drive adoption. This study assesses the factors that determine the probability and intensity of adoption of CSAPs in Bihar, India using cross-sectional data collected in the second half of 2013. The CSAPs considered in this study include site-specific nutrient management (SSNM), crop diversification (CD), minimum tillage (MT) and stress-resistant improved seed (IS) varieties.

In Bihar, almost 90 per cent of the population resides in the rural areas and nearly 80 per cent of them are employed in agriculture. The Indo-Gangetic Plains (IGP) of Bihar cover one of the most productive agricultural areas in India, thus increasing agricultural production in this region is crucial for ensuring national food security (Government of Bihar, 2012). However, agriculture in this region is now increasingly affected by climate variability and climate risks. Northern Bihar is generally a highly flood-prone area, whereas southern Bihar is highly drought-prone (Government of Bihar, 2012). Along with a flat topography, the concentration of rainfall between July and September (about 80 per cent of the total annual rainfall) is the major cause of flooding in northern Bihar. Nearly 74 per cent of the total geographical area of Bihar is flood prone and this constitutes about 17 per cent of the total flood-prone areas of India. However, Bihar also suffers from severe droughts when the summer monsoon lessens causing to less-than-normal rainfall. As Bihar lies at the crossroads of the wet eastern coastal regions and the relatively dry continental region of the western plain, regional variations in rainfall distribution and rainfall variability is much higher. Although the average annual rainfall in Bihar is 1,200 mm, there are considerable variations across the northern and southern areas. Generally, the eastern and northern areas receive 2,000 mm rainfall, whereas it is less than 1,000 mm in the western and south-western parts, making them highly vulnerable to drought. Therefore, vagaries of rainfall, recurrent floods and droughts occur in the same season at the same place, severally affecting agriculture. Among the districts in the IGP, the districts in Bihar are found to be highly vulnerable to climate change (Sehgal *et al.*, 2013). Therefore, attaining sustainable agriculture and reducing the vulnerability of agriculture to climate change through the use of CSAPs is a pre-requisite in this region (Aggarwal *et al.*, 2013; Aryal *et al.*, 2014; Sehgal *et al.*, 2013).

CSAP addresses the intertwined challenges of sustainable farming, food security, and issues of changes in climate (FAO, 2013). The term CSAP encompasses farming practices that sustainably increase productivity, enhance resilience/adaptation, reduce greenhouse gasses (GHG) and help achieve national food security and development goals (FAO, 2013, 2010). Technically, any agricultural practices can be considered CSAP as long as they improve productivity or resource-use efficiency, reduce a community's exposure or vulnerability to climate change, reduce GHG emissions and increase carbon sequestration (Neufeldt *et al.*, 2013). For example, the use of high-yielding and stress-tolerant seed varieties/breeds, and the adoption of improved management practices stabilize and increase farm production even under adverse production conditions. Increased farm production and income enhance farmers' ability to cope with extreme weather events. Similarly, agricultural practices, such as MT, proper management of crop residues and precision NM, enhance sequestration of atmospheric carbon into agro-ecosystems and increase resource-use efficiency, thereby reducing GHG emissions without compromising yield.

Previous studies examine CSAPs' adoption in terms of a single technology/practice and typically fail to account for technology complementarity and substitution possibilities. However, farmers can adopt technologies as complements and substitutes that addresses their multitude of constraints such as moisture stress, low soil fertility, declining groundwater table, terminal heat stress and low crop productivity. If some portfolios of CSAPs are substitutes, agricultural policy needs to concentrate on making these combinations of CSAPs available to farmers. Conversely, if some portfolios of CSAP are complements, it is vital to find ways of offering these as packages because their partial adoption will not achieve the desired productivity or environmental outcomes. Technology adoption decisions can be path dependent: the choice of technologies adopted most recently by farmers may depend on their earlier technology choice. Complementarities among technologies increase income and reduce crop failure substantially, which further stimulates the adoption of multiple technologies (Kassie *et al.*, 2015, 2013; Teklewold *et al.*, 2013; Yu *et al.*, 2012). Recent empirical studies in Africa demonstrate that the joint adoption of these technologies (MT, CDs and ISs) significantly increases net income and reduces production risk compared to the individual adoption of these technologies (Kassie *et al.*, 2015; Teklewold *et al.*, 2013), thus suggesting complementary effects. Therefore, analysis of technology adoption without proper controlling for technology inter-dependence could either underestimate or overestimate the influences of various factors on the decision to adopt (Wu and

Babcock, 1998). Applying multivariate and ordered probit (OP) models, this paper provides some insights on the long-standing discussions on whether farmers adopt CSAPs in a piecemeal or in a composite way. The multivariate and OP models acknowledge the possibility of correlations between adoption decisions across different CSAPs.

The remainder of the paper is outlined as follows: Section 2 deals with the econometrics model and estimation strategies used in the paper. Section 3 provides a brief overview of the study area, data and the variables used in the analysis. Section 4 presents the results and discussion while the last section concludes the study.

2. Econometric model employed for analysis
This study used a multivariate probit (MVP) model to assess the factors affecting the likelihood of adopting multiple CSAPs and an OP model for estimating the level of adoption of CSAPs.

2.1 Multivariate probit model: to identify the determinants of adoption of multiple CSAPs
A technology adoption decision is influenced by multiple factors. An individual farmer may need to adopt a mix of CSAPs to address a multitude of climate risks and agricultural production constraints. Given that many CSAPs are not mutually exclusive, the decision to adopt one of the CSAPs may influence the decision to adopt other CSAPs. Hence, attempting univariate modeling would exclude useful economic information about interdependent and simultaneous adoption decisions (Kassie *et al.*, 2013) – therefore, we applied an MVP model.

MVP models, unlike univariate probit models, allow for potential correlation among the unobserved disturbances in the adoption equations and the relationships between the adoptions of different CSAPs. This means the model considers the possible complementarities (positive correlation) and substitutability (negative correlation) between the CSAPs. Estimation without considering trade-off and synergies of technology adoption leads to biased and inefficient estimates of the determinants of adoption (Greene, 2003). Such biases may lead to situations where researchers may observe the lack of adoption because of poor returns as complementary practices are not adopted. However, one may fail to account for this impact because the univariate models do not adequately correct for these complementarities. For example, many farmers who use MT may also use SSNM; yet, unless researchers analyze this effect, they will not be able to understand the factors that enhance the uptake of SSNM by the farm household.

Theoretically, a specific CSAP is more likely to be adopted if the utility from its adoption is greater compared to its non-adoption. Assume a ith farm household ($i = 1,2,....,N$) confronting a choice on whether to adopt the jth CSAPs (where j represents the choice of SSNM (L), MT (M), CD (D) and stress-resistant IS variety (S)) on its farm plot ($p = 1,....,P$). Let U_0 and U_j denote the benefits to the farm household from the adoption of conventional agricultural practices and CSAPs, respectively. A farm household chooses to adopt the jth CSAPs on its farm plot p if the net benefit (y_{ipj}^*) from the adoption of other available technologies is higher – i.e. $B_{ipj}^* = U_j^* - U_0 > 0$. In this case, the net benefit of the adoption of the CSAPs is a latent variable (i.e. y_{ipj}^*), which is determined by the observed household, farm plot and location information (X_{ip}) and the error term (ε_{ip}) as follows:

$$y_{ipj}^* = X_{ip}'\beta_j + \varepsilon_{ip} \qquad (j = L, M, D, S) \qquad (1)$$

Authors can present equation (1) in terms of the indicator function. In the current study, the unobserved preferences in equation (1) transform into the observed binary outcome equation for each CSAP choice as follows:

$$y_{ipj} = \begin{cases} 1 & \text{if } B^*_{ipj} > 0 \\ 0 & \text{otherwise} \end{cases} \qquad (j = L, M, D, S) \tag{2}$$

In the MVP model, with the prospect of adopting multiple CSAPs, the error terms jointly follow a multivariate normal distribution with zero conditional mean and variance normalized to unity, i.e. $(u_L, u_M, u_D, u_S) \rightarrow^{MVN} (0, \Omega)$ and the covariance matrix (Ω) is given by:

$$\Omega = \begin{bmatrix} 1 & \rho_{LM} & \rho_{LD} & \rho_{LS} \\ \rho_{ML} & 1 & \rho_{MD} & \rho_{MS} \\ \rho_{DL} & \rho_{DM} & 1 & \rho_{DS} \\ \rho_{SL} & \rho_{SM} & \rho_{SD} & 1 \end{bmatrix} \tag{3}$$

where ρ refers to the correlation between error terms – if error terms correlation shown in the off-diagonal elements of the variance-covariance matrix become non-zero, then equation (2) becomes an MVP model.

A pooled MVP model is consistent if the unobserved heterogeneities (ability, motivation, land quality,etc.) are not correlated with observed covariates. The data used in this study afford us a panel structure arising from repeated plot observations per household which authors exploited to estimate equation (2) with the Mundlak approach to control for unobserved heterogeneities (Mundlak, 1978). Mundlak's approach involves including the mean of plot-varying explanatory variables as additional covariates in the regression model. Many studies (Teklewold *et al.*, 2013; Kassie *et al.*, 2013) have applied this approach using cross-sectional multiple plot observations.

2.2 Ordered probit model: for estimating the level of adoption of CSAPs
To analyze the factors determining the intensity of CSAP adoption, the authors applied an OP model. In this case, the dependent variable can take values 0, 1, 2, 3 and 4, depending upon whether a farmer has not used any CSAPs or used at least one, or two or three or four. This is done because the MVP model specified above cannot differentiate how many CSAPs are used by the farmers. When farmers adopt multiple CSAPs, it is difficult to define a cut-off point between adopters and non-adopters, while examining the intensity of CSAP adoption by farmers (Wollni *et al.*, 2010). In our study area, some farmers adopt some CSAPs on part of their land; thus, using a fraction of its area under CSAP as a variable to measure the intensity of adoption (as is often done in other studies) is difficult. To overcome this problem, the authors used the number of CSAPs adopted by farmers as a dependent variable measuring the intensity of adoption and applied an OP model (D'Souza *et al.*, 1993; Teklewold *et al.*, 2013). Poisson regression could have been used assuming that number of CSAPs adopted as a count variable; however, it assumes the equal probability of adoption of each alternative CSAP. In our case, this is not a valid assumption because the likelihood of adopting the first CSAP might differ from the probability of adopting the second and so on. This happens mainly because of the exposure of farmers to information about CSAPs. As the authors have data on the farm plot level, this gives us a possibility to also use both pooled and random effect models and Mundlak's (1978) approach by including the mean of

plot varying covariates that help to capture the correlation between observed covariates and unobserved heterogeneity (Teklewold *et al.*, 2013).

2.3 Concerns in estimations of the econometric models

Multiple factors influence farmers' decisions to adopt agricultural technologies (Birthal *et al.*, 2015; Doss, 2003; Kassie *et al.*, 2010). Hence, it is essential to control for a number of factors in estimating the MVP model. However, when the authors add a number of explanatory/ independent variables to the MVP model, the authors may risk the problem of multicollinearity and the need for a large sample size. The authors applied a condition index to test for the possible multicollinearity (Belsley, 1991; Belsley *et al.*, 2005). The simple decision rule is that if the value of the condition index is below 30, it indicates that there is no severe problem of multicollinearity. Moreover, estimators are projected based on the asymptotic theory and thus, it necessitates a sufficiently large sample size. The number of observations ought to be higher than $1.5k$ ($k + 1$), where k refers to the total number of variables used in the MVP model (Behera *et al.*, 2015; Jöreskog and Sörbom, 1993). If the number of observations is less than an obligatory number, the asymptotic variance-covariance matrix is unlikely to be positive definite. This is exhibited in biased inference caused by poor estimates of parameter variance-covariance.

3. Study area, data and description of variables

The data for this study were collected in 2013 from 641 farm households residing in 12 villages of Vaishali district of Bihar state, India (Table I and Figure 1). A multi-stage sampling method was applied to select the sample households in the study. In the first stage, study villages within the district were chosen using a stratified random sampling method. In the second stage, a census of around 75 per cent of households in the village was taken to collect basic information, such as main occupation, crops grown, operational land holdings and age and gender of the household head. Finally, based on the information gathered in the village census, the sample households were randomly selected from each village. The authors collected information that includes household characteristics, farm plot characteristics, access to credit, extension services, market characteristics and information on training received, adoption of CSAPs (including MT), CD and stress-resistant IS varieties.

Vaishali district lies in the Eastern IGP. It has a sub-humid tropical climate with average annual rainfall of about 1150 mm, 80 per cent of which is received between July and September. Although rice and wheat are the major crops, maize is emerging as an

Table I. Distribution of sample households by village in Vaishali district, Bihar

Village name	Sample size
Bhatha Dasi	63
Bilandpur	68
Dedhpur	46
Dhabhaich	46
Laxminarayanpur	44
Mirpur	55
Mukundpur	69
Panapur Camp	56
Raja Pakar	70
Rampur Ratnagar	45
Rasalpour	48
Varishpur	31

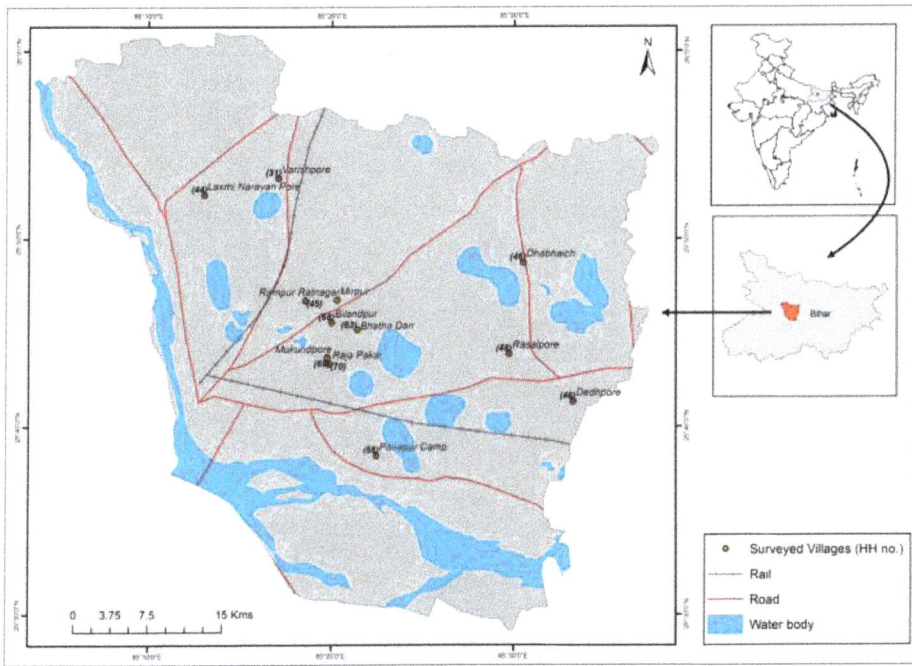

Figure 1. Map of the study area

alternative crop in this area. The average size of land holdings in Vaishali district is 0.51 ha. Land fragmentation is a critical issue in Vaishali with an average farm plot size of 0.12 ha.

3.1 Description of variables used in the study and hypotheses to be tested
The descriptive statistics and explanation of the variables (both dependent and explanatory) employed in the study are presented in Table II.

3.1.1 Dependent/explained variables. CD (legume integration in crop rotations and intercropping) is one of the CSAPs under study. About 20 per cent of the total farm-plots under study follow CD. Such practices help farmers to diversify to reduce risk and increase productivity and income through improving soil fertility, controlling for pests and weed infestation (Bradshaw *et al.*, 2004; Liebman and Dyck, 1993; Lin, 2011; Shiferaw *et al.*, 2012; Teklewold *et al.*, 2013; Kassie *et al.*, 2015).

The adoption of improved technologies, such as improved and stress-tolerant varieties, boosts farmers' adaptive and resilience capacities. Farmers mainly use drought-tolerant varieties of rice such as Swarna Sub 1, and in some cases were found to have used submergence-tolerant varieties. In addition to increasing income through improving productivity, stress-tolerant varieties can stabilize productivity through minimizing the effects of climate change and variability (Sarkar *et al.*, 2006).

MT, which refers to either reduced tillage or zero tillage with residue retention, is another CSAP under study. It entails minimum soil disturbance and allowing crop residue or stubble to remain on the plot with the accompanying benefits of better soil aeration, carbon sequestration, improved soil fertility and an increased water-holding capacity of soil (Lal, 2004, 1997; Sapkota *et al.*, 2015). It can cope with water-stress situations by conserving

Table II. Description of variables used in the study

Variables	Mean	SD	Variable Description
Dependent Variables			
CD (D)	0.20	0.40	1 if crop diversification is practiced and 0 if not
IS (D)	0.03	0.17	1 if stress tolerant seed variety used and 0 if not
MT (D)	0.19	0.39	1 if MT; 0 if conventional tillage
NM (D)	0.18	0.39	1 if SSNM and 0 if not
Total CSAPs	0.43	0.52	Adoption intensity; integer value from 0 to 4
Independent variables			
Household characteristics			
Male headed household (D)	0.92	0.27	1 if male and 0 if female
General caste (D)	0.33	0.47	1 if general caste and 0 otherwise
Age (C)	51	13	Age of household head in year
Literate HH head (D)	0.64	0.48	1 if literate and 0 otherwise
Literate spouse of HH head (D)	0.27	0.44	1 if literate and 0 otherwise
Family size (C)	6.40	2.88	Number of family members
Migrant (D)	0.32	0.47	1 if at least one migrant member and 0 otherwise
Farmland characteristics			
Tenure of plot (D)	0.76	0.43	1 if owned and 0 if rented in
Area of plot (C)	0.12	0.13	Area of plot in hectare
Soil fertility (D)	0.46	0.50	1 if good and 0 otherwise
Soil depth (D)	0.15	0.36	1 if deep and 0 if shallow
Land slope (D)	0.47	0.50	1 if gentle slope and 0 if medium/steep slope
Distance (C)	0.75	1.05	Distance from house to plot in Km
Economic and social capital			
Land operated (in ha) (C)	0.51	0.52	Total land operated
TLU (C)	0.70	0.82	Livestock owned
Asset index	0.30	0.53	Household asset index
Credit access (D)	0.39	0.49	1 if farmer has credit access and 0 otherwise
Association in group (D)	0.11	0.31	1 if membership in any association and 0 otherwise
Access to market and agricultural extension service and training			
Distance to local market (C)	2.46	1.49	Distance to local market from house (in km)
Distance to agriculture extension service (C)	4.67	3.71	Distance to agriculture extension service from house (in km)
Training (D)	0.24	0.41	1 if participated in agricultural training and 0 otherwise
Main source of information			
Farmer to farmer communication (D)	0.46	0.50	1 if information received from neighbor/relative farmer/cooperative and 0 otherwise
Government extension (D)	0.03	0.12	1 if information received from government extension service and 0 otherwise

(continued)

Table II.

Variables	Mean	SD	Variable Description
ICT (D)	0.03	0.17	1 if information received from radio, newspaper, TV, mobile and 0 otherwise
Seed traders/private company (D)	0.28	0.45	1 if information received from seed traders/private company and 0 otherwise
Climate risks experienced by household over the past five years			
Heat stress (D)	0.72	0.45	1 if farmer experienced heat stress and 0 otherwise
Less rainfall (D)	0.91	0.28	1 if farmer experienced less rainfall and 0 otherwise
Decrease winter (D)	0.27	0.44	1 if farmer experienced decrease winter and 0 otherwise
Total number of HH (plot)	641(2004)		

Notes: C – Continuous variable; D-Dummy variable

moisture and utilizing residual moisture during planting, and with temperature-stress by advancing the planting time of winter crops after the harvest of summer crops. About 19 per cent of the farm plots operated by the sample farmers are under MT.

SSNM contributes to the management of nutrients needed for crops, thereby improving resource-use efficiency and lowering GHG emissions from agriculture (Sapkota *et al.*, 2014). About 18 per cent of the total farm-plots received SSNM.

The dependent variable, total CSAPs, refers to the number of CSAPs used by the farmers in the same plot in 2012. This number ranges from 0 to 4 in our case.

3.1.2 Independent/explanatory variables. The authors specified the model based on past theoretical frameworks and empirical adoption literature (Aryal *et al.*, 2014, Aryal and Holden, 2011; Erenstein and Farooq, 2009; Feder and Umali, 1993; Kassam *et al.*, 2009; Kassie *et al.*, 2013, 2010, 2015; Pender and Kerr, 1998; Teklewold *et al.*, 2013). Description of the explanatory variables and hypothesis about their effects on the dependent variable(s) are discussed below.

3.1.2.1 Household characteristics. A household's socio-economic and demographic characteristics consist of the main attributes of the head of the household, such as education, literacy status, age and gender, family size, spouse's literacy status and migration status. Household characteristics generally stimulate technology adoption decisions under market imperfection and institutional failure (de Janvry *et al.*, 1991; Holden *et al.*, 2001). Literate household heads have better capability and knowledge to access and absorb new information, and are more likely to have more non-farm income, which in turn influence the decision to adopt new technology (Chander and Thangavelu, 2004). Generally, agricultural technology adoption can be a part of an overall household strategy and thus, the spouse's literacy status might also affect it. The age of the household head is worth examining as older people have more experience with farming systems and often a greater accumulation of physical and social capital; however, they also often have short-term planning horizons, a loss of energy and more risk aversion (Albert and Duffy, 2012). Migration, which refers here to families with at least one member currently living elsewhere, reduces household labor endowment and increases access to alternative income sources. Therefore, it is assumed that migration enables the adoption of new labor-saving technologies.

3.1.2.2 Farmland characteristics. Farm plot features (farm plot size, tenancy status, soil fertility, soil depth, plot slope, and distance to plot from homestead) are included in the analysis.

The soil fertility, soil depth, plot slope, and distance to the plot are provided separately for different parcels (plot by plot). This is crucial to control for the possible impacts of farmland attributes on technology adoption. For instance, distant plots not only require more cost for transporting inputs but are also more difficult to monitor. Thus, farmers may be less keen to embrace new technology on distant plots. Land fragmentation is extremely high and farmers run multiple small plots in Bihar. Small farm plot size can be a limiting factor to mechanization and may make farmers less likely to adopt MT on such plots.

3.1.2.3 Economic and social capital. To control for wealth differences among households, the authors included several economic capital variables such as farm size, livestock ownership, household asset index and labor force available in the household in the analysis. For social capital, the authors used variables such as membership in an organization (e.g. farm cooperatives) or any other farm clubs/input traders/sellers and other groups. The authors used caste as one of the social capitals, signifying variables, as it affects access to public domains in rural communities in South Asia (Aryal and Holden, 2013, 2012). Caste restricts or facilitates a household's participation in some markets and access to information (Yamano et al., 2015). The authors hypothesize that the general caste group is more likely to adopt CSAPs than the backward and scheduled caste groups.

3.1.2.4 Markets and institutional services. Access to markets and other institutional services are important variables, as they influence transaction costs. Distance to village markets is used as a proxy for market access, while the distance to extension services is considered a proxy for access to institutional services.

3.1.2.5 Source of information and training. Adoption of CSAPs also depends on the access to information and training received. Farmers receive information through multiple sources including farmer-to-farmer communication, extension services, information and communication technology and private traders. As the training on related topics such as soil-water management and MT, also influences the farmers' likelihood of adopting those technologies, the authors controlled for this in the analysis.

3.1.2.6 Climate risks. Agriculture faces numerous risks arising from changes in climate. Therefore, farmers embrace new farming practices to adapt to those climate risks. Farmers also stated that heat stress, less rainfall, and a decrease in winter are three leading climate risks experienced during the last five years. These three major climate risks were included in the analysis to investigate the impacts of these shocks on the adoption of CSAPs.

Village dummies are included in the regression to control for spatial differences such as rainfall, infrastructure and quality of service delivery.

4. Results and discussion

4.1 Determinants of multiple adoptions of CSAP: Multivariate probit model

Farmers adopted different CSAPs concurrently; this shows that there is a chance of a correlation between their choices of CSAPs. This is tested using pair-wise correlation coefficients across the residuals of an MVP model. Most of the estimated pair-wise correlation coefficients across the residuals of MVP model are statistically significant (Table III), thereby supporting our hypothesis that the error terms of the multiple decision equations are correlated. The likelihood ratio test [chi2(6) = 78.48; Prob > chi2 = 0.000] also rejected the null hypothesis that the covariance of the error terms across equations are not correlated. This justifies the rationale for using the MVP model and confirms that adoption of multiple CSAPs in Bihar is not mutually exclusive. Practicing CD and MT are significantly and negatively associated (Table III); implying that farmers consider these CSAPs substitutes. Other CSAP combinations such as MT and IS, MT and SSNM and IS and SSNM are significantly and positively associated, implying that farmers consider these as complements (i.e. farmers apply them simultaneously).

Table IV shows the results of the MVP model estimated using the maximum likelihood method on farm plot-level observations. Although the authors have 2,004 farm plots, the authors used only 1,926 farm plot-level observations because of inconsistencies in the remaining observations. Our estimates show that the model fits the data well as the Wald test [Wald chi2(172) = 1362.16; Prob > chi2 = 0.000)] of the null hypothesis, that all regression coefficients in each equation are jointly equal to zero, is rejected. This shows the relevance of the model to account for the unobserved correlations across decisions to adopt multiple CSAPs.

Results show that the explanatory variables affecting decisions to adopt differ substantially across the CSAPs. Unlike other studies, the authors found that male-headed households are less likely to adopt CD and IS. For example, when compared to female-headed households, male-headed households are 0.42 per cent less likely to adopt CD. Compared to the backward/scheduled caste group, farmers belonging to the general caste group are more likely to adopt IS and NM; whereas, they are less likely to adopt MT. These results corroborate the findings by Aryal and Holden (2011) and Yamano et al., (2015) that caste position affects farmers' investment decisions. Older household heads are less likely to adopt CD and NM; the former may be because of a higher requirement of labor, while the latter may be attributable to the fact that older household heads are less familiar with this relatively newer technology. Older household heads are more likely to adopt MT and IS; however, the result is significant at the 10 per cent level only. Literacy status of the household head and spouse are an important factor influencing the CSAP adoption decision. Households with a literate household head are more likely to adopt MT, IS and NM while those with literate spouses are less likely to adopt CD but more likely to adopt MT and NM, indicating that they are more likely to choose labor saving CSAPs. Migration is significantly and negatively associated with the likelihood of adopting NM and IS with 5 and 1 per cent levels of significance, respectively, while it is positively associated with the adoption of CD at the 10 per cent level of significance.

Several farm land characteristics are found to have influenced the probability of adopting CSAPs. Land tenure affects adoption of CD and MT; owner-cultivated plots are less likely to adopt CD and MT. This result is consistent with earlier studies on technology adoption (i.e. Feder and Umali, 1993; Kassie et al., 2013, 2010). In Bihar, tenants need to seek permission from the landlords to grow crops on the rented plots, and thus, CD is less likely in such cases. Another important issue that inhibits mechanization in Bihar is land fragmentation. The authors controlled for this using the area of the farm plots. Our results show that the area of the plot does not affect CD, IS and NM while it affects the adoption of MT significantly and positively. This justifies our hypothesis that larger plots are more

Table III. Correlation of error terms and likelihood ratio tests

Correlation pairs	Coefficient	Standard error
Crop diversification × MT	−0.267***	0.061
Crop diversification × IS	−0.139	0.169
Crop diversification × SSNM	−0.330***	0.058
MT × IS	0.268***	0.101
MT × SSNM	0.289***	0.056
IS × SSNM	0.182**	0.088

Notes: LR test of: rho21 = rho31 = rho41 = rho32 = rho42 = rho43 = 0: chi2(6) = 78.48; Prob > chi2 = 0.000; Significance level: 5% level **; 1% level ***

Table IV. Estimates of the MVP model

Dependent variables	CD	MT	IS	NM
Household (HH) characteristics				
Male headed HH	−0.421** (0.176)	0.024 (0.200)	−1.469*** (0.361)	−0.294 (0.182)
General caste	0.062 (0.106)	−0.441*** (0.106)	0.576*** (0.223)	0.121*** (0.017)
Age of HH head	−0.013** (0.006)	0.006* (0.004)	0.015* (0.008)	−0.005** (0.002)
Literate HH head	0.167 (0.111)	0.054** (0.019)	0.703*** (0.250)	0.212* (0.121)
Literate spouse	−0.402*** (0.121)	0.393*** (0.112)	−0.198 (0.202)	0.294** (0.117)
Family size	−0.003 (0.015)	−0.003 (0.016)	−0.069** (0.034)	0.017 (0.016)
Migrated	0.157* (0.094)	0.111 (0.101)	−0.700*** (0.245)	−0.144** (0.080)
Farm land characteristics				
Tenure of plot	−0.132*** (0.012)	−0.087*** (0.016)	0.121 (0.220)	0.098 (0.113)
Area of plot	−0.048 (0.334)	0.112*** (0.036)	0.746 (0.597)	−0.278 (0.385)
Soil fertility	0.369*** (0.094)	−0.470*** (0.101)	−0.698*** (0.210)	0.047 (0.101)
Soil depth	0.237** (0.116)	−0.610*** (0.167)	−0.822 (0.518)	−0.465*** (0.149)
Land slope	−0.234** (0.091)	0.255*** (0.099)	−0.284 (0.211)	0.646*** (0.099)
Distance to plot	−0.006 (0.041)	−0.125*** (0.051)	0.063 (0.075)	−0.035** (0.017)
Economic and social capital				
Land operated	0.264** (0.104)	0.163*** (0.011)	0.067*** (0.023)	0.136** (0.062)
Livestock owned	0.215*** (0.053)	0.198*** (0.059)	0.226* (0.138)	0.205*** (0.057)
Asset index	0.343*** (0.133)	0.275** (0.132)	0.336** (0.171)	0.256** (0.119)
Credit access	−0.537*** (0.102)	0.575*** (0.113)	−0.337 (0.260)	0.680*** (0.113)
Association in group	0.297** (0.143)	0.394** (0.188)	−0.411 (0.364)	0.537*** (0.184)
Access to market and agricultural extension services and training				
Distance to market	−0.035 (0.034)	−0.112*** (0.040)	−0.080 (0.083)	−0.087** (0.040)
Distance to extension service	−0.061*** (0.015)	−0.045*** (0.013)	−0.071*** (0.024)	0.002 (0.014)
Training	0.645* (0.359)	0.162** (0.073)	0.115** (0.052)	0.058** (0.029)
Source of information				
Farmer-to-farmer communication	0.244*** (0.097)	0.561*** (0.112)	0.914*** (0.287)	0.932*** (0.125)
Government extension service	0.838*** (0.320)	0.084** (0.037)	−0.179 (0.327)	0.169 (0.333)
ICT	0.818** (0.401)	0.516* (0.307)	0.169*** (0.039)	−0.413 (0.292)
Private traders	−0.156 (0.108)	0.127** (0.063)	0.167** (0.075)	0.264** (0.107)
Major climate risks				
High temperature	0.639*** (0.113)	0.504*** (0.108)	0.482** (0.231)	0.081 (0.107)
Decreasing rainfall	0.183 (0.165)	0.064 (0.154)	0.143*** (0.056)	−0.159 (0.161)
Short winters	0.051 (0.115)	1.378*** (0.152)	−0.180 (0.228)	−0.605*** (0.135)
Joint significance of village dummies				
$Chi^2(11)$	74.41	84.62	63.11	222.19
$Prob > Chi^2(11)$	0.000	0.000	0.000	0.000
Constant	−1.920*** (0.395)	1.114*** (0.391)	−1.694*** (0.734)	0.378** (0.167)
Number of observations	1926	1926	1926	1926

Notes: LR test of: rho21 = rho31 = rho41 = rho32 = rho42 = rho43 = 0: chi2(6) = 78.48 Prob > chi2 = 0.000; Wald chi2(156) = 1362.16; Prob > chi2 = 0.000; Log likelihood= −1916.92; Significance level: 10% level (*); 5% level **; 1% level ***

likely to receive MT as it requires the operation of machines on the plot which is more difficult on small plots. Farmers are more likely to adopt CD on plots with deep and fertile soil. Contrary to our hypothesis, farmers are less likely to adopt MT and IS on plots with deep soil. Consistent with earlier studies (Feder and Umali, 1993; Kassie et al., 2013, 2010), our results show that farmers are less likely to adopt MT and IS on the plots when the soil quality appears to be good (i.e. plots with deep soil depth).

The economic and social capital of the farm household affects their decisions to adopt technology. Our results show that farmers of large farms are more likely to adopt all CSAPs under study. For instance, larger farm size is positively associated with the adoption of MT and IS; the coefficients are significant at the 1 per cent level. Households with more livestock ownership have a higher probability of adopting CD, MT, IS and NM. Access to credit has variable effects on technology adoption: it increased the probability of adopting MT and NM while it decreased the probability of CD and IS. Association in village cooperatives and groups is more likely to increase the adoption of CD, MT and NM.

Distance to market from the household is negatively and significantly associated with the adoption of MT and NM; however, it has no significant association with other CSAPs. This is plausible because MT requires machines, which are mostly used on a custom hire basis and are available only in the market center in Bihar. Households that are far from extension services are less likely to adopt CD, MT and IS.

Farmers obtain information about the technologies and farming practices from different sources such as other farmers, government extension services, information and communication technologies (ICTs) (i.e. information through mobile phones and private traders. The decisions to adopt different technologies are affected by the type of information sources. As per our *a priori* assumption, the results show that farmer-to-farmer communication is positively related to the adoption of CD (significant at the 1 per cent level). Information from government extension services increases the probability of adopting CD and MT, while it does not affect other CSAP adoptions. This is possible because government extension agents are less informed about the relatively newer CSAPs in Bihar, though addressing the impact of climate change on agriculture is a matter of serious concern. ICTs (especially the use of mobile phones to provide information on agricultural technology and weather information) are becoming a popular method of communication in India (Birthal et al., 2015; Rao, 2007). Use of ICTs is associated positively and significantly with the adoption of CD, MT and IS. Private traders are also an important source of information about CSAPs now. Farmers receiving information from private traders are more likely to adopt MT, IS and NM. This is plausible in Bihar, as the majority of traders are linked to major products such as rice, wheat, and maize; thus, they focus on providing information that expands their businesses. Participation in agricultural training is positively and significantly associated with the likelihood of adopting CSAPs, especially CD, MT, IS and NM.

Farmers' choice of CSAPs also depends on their experiences regarding climate risks. Farmers experiencing high temperatures as major climate risks over the past five years are more likely to adopt CD, MT, and IS. Farmers are adopting IS (especially stress-tolerant varieties, such as Swarna sub1) as a result of experiencing climate risks, such as heat stress because of high temperature (Yamano et al., 2015). Another climate risk, decreasing rainfall, is significantly and positively associated with the adoption of IS. Farmers facing short winters are more likely to adopt MT and less likely to adopt NM. One of the main reasons behind this could be the use of the zero tillage (in our case the authors call it MT) wheat production system, in which farmers can plant wheat earlier than they do conventionally and thereby the wheat escapes the terminal heat if the winter is short. (Sapkota et al., 2015).

4.2 Factors explaining intensity of CSAPs adoption

Farmers have adopted multiple CSAPs in the study area; however, the intensity of adoption varies from 1 to 4 CSAPs in the same farm plot by the sampled farm households. The authors estimated the OP model to examine the factors explaining the intensity of CSAPs adoption. As a direct interpretation of the coefficients of an OP model is less informative, the authors calculated marginal effects on each outcome (i.e. level of intensity [Table V]). The chi^2 statistic for the OP model is statistically significant at less than the 1 per cent level of significance. This indicates that the null hypothesis – that joint test of all slope coefficients equal zero – is rejected.

Results show that several factors influence the intensity of CSAP adoption. Gender of the household head, caste, literacy status of household head and migration are the major household characteristics variables affecting the intensity of CSAP adoption. The number of CSAPs adopted is significantly lower among the male-headed households compared to female-headed households. Unlike the adoption decision (as shown in Table IV), the variable "literate spouse" is found to have no effect on the intensity of CSAPs used (Table V). This indicates how the same variable can have a differing impact on the decision to adopt CSAPs and the intensity of adoption. Households with at least one migrated member have a significant and positive effect on the number of CSAPs used, with varying marginal probabilities of each outcome.

Compared to the decision to adopt CSAPs (Table IV), only three farm land characteristics variables including tenure status, the size of and distance to the plot, are found to have a significant impact on the intensity of CSAP adoption. Owner-cultivated plots have a higher intensity of CSAPs used than rented plots (significant at the 1 per cent level). Larger plots have a positive association with the intensity of adoption, indicating that land fragmentation in Bihar is a constraining factor to CSAP adoption. Plots farther from the homestead received fewer CSAPs compared to plots nearer the homestead. This could be because of the higher transportation and monitoring costs to farmers.

Wealthier households (households with higher asset indices) are found to have a higher intensity of adopting CSAPs. Association in groups increases the level of CSAPs used positively and significantly. Credit access has a negative and significant effect on CSAPs used. This is plausible in the study area as credit taken for agricultural purposes is often used for other social purposes. The focus group discussion with the farmers in the study area revealed that in several cases, credit is used to cover marriage, dowry and medical expenses. Access to markets and agricultural extension services are crucial factors to enhance the level of CSAP adoption. Our results show that the variables of increasing distance from markets and extension services have negative and significant effects on the intensity of CSAPs used.

Information sources, including "farmer-to-farmer communication", "government extension services" and "information and communication technology" are positively and significantly associated with the level of CSAP adoption. This indicates the importance of farmers' access to these information sources and mobilizing resources to upgrade these facilities (Birthal *et al.*, 2015). Moreover, this indicates that national and international agricultural institutions are effective in enhancing farmers' knowledge about the CSAPs required to address the climate risks in agriculture. Training on soil management and tillage, and CD has a positive and significant effect on the level of CSAPs used.

Farmers who faced high temperatures as a major climate risk over the past five years have significantly higher levels of CSAP adoption, opposed to the farmers who faced short winters as the major climate risk over the last five years.

Table V. Estimates of the OP model and the marginal effects

CSA intensity	Coefficient	Pr(Y = 0|X)	Marginal effects on each outcome Pr(Y = 1|X)	Pr(Y = 2|X)	Pr(Y = 3|X)
Household (HH) characteristics					
Male headed HH	−0.365*** (0.139)	0.144*** (0.055)	−0.137*** (0.052)	−0.007** (0.003)	−0.005 (0.013)
General caste	0.097*** (0.031)	−0.057* (0.031)	0.036* (0.020)	0.013*** (0.001)	0.002** (0.001)
Age of HH head	0.002 (0.003)	−0.001 (0.001)	0.001 (0.001)	0.000 (0.000)	0.000 (0.000)
Literate HH head	0.239*** (0.087)	−0.090*** (0.032)	0.088*** (0.032)	0.013** (0.006)	0.003 (0.002)
Literate spouse	0.012 (0.086)	−0.005 (0.033)	0.005 (0.032)	0.001 (0.001)	−0.001 (0.001)
Family size	−0.012 (0.012)	0.005 (0.005)	−0.005 (0.004)	−0.000 (0.000)	−0.001 (0.001)
Migrant	0.086*** (0.033)	−0.083*** (0.028)	0.032** (0.016)	0.021* (0.011)	0.002** (0.001)
Farmland characteristics					
Tenure of plot	0.134*** (0.050)	−0.052* (0.031)	0.050* (0.030)	−0.004*** (0.001)	0.002** (0.001)
Area of plot	0.131** (0.064)	−0.150* (0.102)	0.049*** (0.019)	0.006** (0.003)	0.001 (0.003)
Soil fertility	−0.096 (0.074)	0.037 (0.028)	−0.036 (0.027)	−0.017 (0.014)	−0.001 (0.001)
Soil depth	−0.134 (0.098)	0.050 (0.036)	−0.049 (0.036)	−0.001 (0.001)	−0.002 (0.002)
Land slope	−0.032 (0.071)	0.012 (0.027)	−0.012 (0.026)	−0.000 (0.001)	−0.001 (0.001)
Distance to plot	−0.128*** (0.033)	0.031** (0.012)	−0.024** (0.012)	−0.004 (0.002)	0.002** (0.001)
Economic and social capital					
Land operated	0.231*** (0.080)	−0.088*** (0.031)	0.086*** (0.030)	0.003** (0.001)	0.003*** (0.001)
Livestock owned	0.203*** (0.043)	−0.078*** (0.016)	0.075*** (0.016)	0.009*** (0.001)	0.002** (0.001)
Asset index	0.362*** (0.097)	−0.142*** (0.038)	0.136*** (0.036)	0.006** (0.002)	0.004** (0.002)
Credit access	−0.685*** (0.078)	0.251*** (0.026)	−0.244*** (0.026)	−0.007*** (0.002)	−0.004** (0.002)
Association in group	0.112*** (0.036)	−0.073** (0.045)	0.042** (0.019)	0.011** (0.005)	0.003 (0.002)
Access to market and agricultural extension services, and training					
Distance to market	−0.063** (0.027)	0.024** (0.010)	−0.023** (0.010)	−0.001** (0.000)	−0.003* (0.002)
Distance to extension service	−0.061*** (0.010)	0.024*** (0.004)	−0.023*** (0.004)	−0.001*** (0.000)	−0.003*** (0.001)
Training	0.628*** (0.094)	−0.244*** (0.036)	0.232*** (0.034)	0.013*** (0.003)	0.006*** (0.002)
Source of information					
Farmer to farmer communication	0.322*** (0.088)	−0.197*** (0.032)	0.191*** (0.031)	0.006*** (0.002)	0.002** (0.001)
Government extension service	0.121*** (0.299)	−0.297*** (0.054)	0.293*** (0.054)	0.004*** (0.001)	−0.003*** (0.001)
ICT	0.225*** (0.281)	−0.333*** (0.040)	0.328*** (0.040)	0.004*** (0.001)	−0.005*** (0.001)
Private traders	−0.019 (0.078)	0.007 (0.030)	−0.007 (0.029)	−0.001 (0.001)	0.000 (0.001)

(continued)

Table V.

CSA intensity	Coefficient	Marginal effects on each outcome			
		Pr(Y = 0|X)	Pr(Y = 1|X)	Pr(Y = 2|X)	Pr(Y = 3|X)
Major climate risks					
High Temperature	0.898*** (0.084)	−0.309*** (0.025)	0.301*** (0.024)	0.017*** (0.004)	0.008*** (0.002)
Decreasing rainfall	0.061 (0.119)	−0.023 (0.045)	0.023 (0.043)	0.001 (0.001)	0.001 (0.001)
Short winters	−0.682*** (0.090)	0.243*** (0.029)	−0.237*** (0.028)	−0.016*** (0.005)	−0.007*** (0.002)
Joint significance of village dummies					
Chi2(11)	46.30				
Prob > Chi2	0.000				
cut1_cons	1.301*** (0.434)				
cut2_cons	2.663*** (0.440)				
cut3_cons	1.234*** (0.123)				
Pseudo R-Squared	0.274				
Log likelihood	−1116.189				
Number of observations	1926	1926	1926	1926	1926

Notes: *, ** and ***indicate 10, 5 and 1% significance levels, respectively. Figures in parentheses are standard errors

5. Conclusions

This study assessed the factors that determine the likelihood and intensity of adoption of CSAPs by the farmers of Bihar, India. Understanding barriers to and enabling conditions for the adoption of CSAPs helps in designing and formulating extension messages and agricultural policies that can accelerate the dissemination of CSAPs. Our results show that farmers adopt these CSAPs as complements and substitutes; however, there is greater scope for promoting complementarities among these CSAPs. Farmers' characteristics including gender, caste, education, social and economic capital, farm land characteristics, access to the market, extension services and training, and major climate risks experienced by the local farmers, are found to be some of the major factors affecting the decision to adopt the technology.

Access to markets and extension services is found to have a crucial role in increasing the uptake of CSAPs. Therefore, it is important to focus on policies and plans that improve market access and quality of public extension services. More training on CSAPs for farmers, more government extension staff working at the local level and the use of ICTs to share and promote knowledge of CSAPs are essential. In this case, mainstreaming CSAPs in village-level (local) climate-change adaptation plans and disseminating knowledge of new CSAPs to rural farmers through enhanced extension services is crucial. Another important issue of high concern is the increasing land fragmentation in Bihar and its implications on the adoption of CSAPs. Given that farmers operating larger farms are more likely to adopt CSAPs; it is important to facilitate agricultural land rental markets, mainly because acquiring land through land sale markets is beyond the capacity of the poor farmers. Therefore, agricultural policies should focus on the regulation of agricultural land rental markets in such a way that landlords do not need to compromise their property rights while renting land and, at the same time, efficient small farmers can access land through well-regulated land rental transactions.

Acknowledgements

The authors acknowledge the support of the CGIAR research program on Climate Change, Agriculture and Food Security (CCAFS) and CGIAR Research Program on Wheat Agri-Food Systems (CRP-WHEAT) for this study. They thank also to all field staff for their sincere efforts while collecting data in Bihar. Authors also sincerely thank Mr Amit Kumar Srivastava for his assistance in providing an excellent map of the study area. The views expressed here are those of the authors and do not necessarily reflect the views of the authors' institutions or CCAFS/CRP WHEAT. The usual disclaimer applies.

References

Aggarwal, P.K., Bhatta, G.D., Joshi, P.K., Prathapar, S.A., Jat, M.L. and Kadian, M. (2013), *Climate Smart Villages in South Asia*, Climate Smart Agriculture Learning Platform, South Asia, New Delhi.

Albert, S.M. and Duffy, J. (2012), "Differences in risk aversion between young and older adults", *Neuroscience and Neuroeconomics*, Vol. 2012 No. 1, pp. 1-12.

Aryal, J.P. and Holden, S.T. (2011), "Caste, investment and intensity of production", *18th Annual Conference, European Association of Environmental and Resource Economists (EAERE)*, 29 June - 2 July, *Rome*.

Aryal, J.P. and Holden, S.T. (2012), "Livestock and land share contracts in a Hindu society", *Agricultural Economics*, Vol. 43 No. 5, pp. 593-606.

Aryal, J.P. and Holden, S.T. (2013), "Land reforms, caste discrimination and land market performance in Nepal", in Holden, S., Otsuka, K. and Deininger, K. (Eds), *Land Tenure Reform in Asia and Africa*, Palgrave Macmillan, Basingstoke, p. 25.

Aryal, J.P., Sapkota, T.B., Jat, M.L. and Bishnoi, D.K. (2015), "On-farm economic and environmental impact of zero-Tillage Wheat: a case of North-west India", *Experimental Agriculture*, Vol. 51 No. 1, pp. 1-16.

Aryal, J.P., Farnworth, C.R., Khurana, R., Ray, S. and Sapkota, T.B. (2014), "Gender dimensions of climate change adaptation through climate smart agricultural practices in India", *Innovation in Indian Agriculture: Ways Forward,* Institute of Economic Growth (IEG), New Delhi, and International Food Policy Research Institute (IFPRI), *New Delhi and Washington, DC.*

Aryal, J.P., Sapkota, T.B., Stirling, C.M., Jat, M.L., Jat, H.S., Rai, M., Mittal, S. and Sutaliya, J.M. (2016), "Conservation agriculture-based wheat production better copes with extreme climate events than conventional tillage-based systems: a case of untimely excess rainfall in Haryana, India", *Agriculture, Ecosystems & Environment*, Vol. 233, pp. 325-335.

Behera, B., Rahut, D.B., Aryal, J. and Ali, A. (2015), "Household collection and use of biomass energy sources in South Asia", *Energy*, Vol. 85, pp. 468-480.

Belsley, D.A. (1991), "A Guide to using the collinearity diagnostics", *Computer Science in Economics and Management*, Vol. 4 No. 1, pp. 33-50.

Belsley, D.A., Kuh, E. and Welsch, R.E. (2005), *Regression Diagnostics: Identifying Influential Data and Sources of Collinearity*, John Wiley & Sons, Hoboken, New Jersey.

Birthal, P.S., Kumar, S., Negi, D.S. and Roy, D. (2015), "The impacts of information on returns from farming: evidence from a nationally representative farm survey in India", *Agricultural Economics*, Vol. 46 No. 4, pp. 549-561.

Bradshaw, B., Dolan, H. and Smit, B. (2004), "Farm-level adaptation to climatic variability and change: crop diversification in the Canadian Prairies", *Climatic Change*, Vol. 67 No. 1, pp. 119-141.

Chander, P. and Thangavelu, S.M. (2004), "Technology adoption, education and immigration policy", *Journal of Development Economics*, Vol. 75 No. 1, pp. 79-94.

D'Souza, G., Cyphers, D. and Phipps, T. (1993), "Factors affecting the adoption of sustainable agricultural practices", *Agricultural and Resource Economics Review*, Vol. 22 No. 2, pp. 159-165.

de Janvry, A., Fafchamps, M. and Sadoulet, E. (1991), "Peasant household behaviour with missing markets: some paradoxes explained", *The Economic Journal*, Vol. 101 No. 409, pp. 1400-1417.

Doss, C.R. (2003), *Understanding Farm Level Technology Adoption: lessons Learned from CIMMYT's Micro Surveys in Eastern Africa*, CIMMYT, Mexico, DF.

Erenstein, O. and Farooq, U. (2009), "A survey of factors associated with the adoption of zero tillage wheat in the irrigation plains of southAsia", *Experimental Agriculture*, Vol. 45 No. 2, pp. 133-147.

Erenstein, O., Farooq, U., Malik, R.K. and Sharif, M. (2008), "On-farm impacts of zero tillage wheat in South Asia's rice–wheat systems", *Field Crops Research*, Vol. 105 No. 3, pp. 240-252.

FAO (2010), *Climate-Smart Agriculture: Policies, Practices and Financing for Food Security, Adaptation and Mitigation*, Food and Agriculture Organization of the United Nations, Rome.

FAO (2013), *Climate-Smart Agriculture: Sourcebook*, Food and Agriculture Organization of the United Nations, Rome.

Feder, G. and Umali, D.L. (1993), "The adoption of agricultural innovations: a review", *Technological Forecasting and Social Change*, Vol. 43 Nos 3/4, pp. 215-239.

Government of Bihar (2012), *State Action Plan on Climate Change: Building Resilience through Development*, Government of Bihar, Bihar.

Greene, W.H. (2003), *Econometric Analysis, Journal of the American Statistical Association*, Education, P. (Ed.), Prentice Hall, Upper Saddle River, NJ.

Holden, S., Shiferaw, B. and Pender, J. (2001), "Market imperfections and land productivity in the Ethiopian highlands", *Journal of Agricultural Economics*, Vol. 52 No. 3, pp. 53-70.

Jöreskog, K.G. and Sörbom, D. (1993), *LISREL 8: Structural Equation Modeling with the SIMPLIS Command Language*, Scientific Software International, Chicago, IL.

Kassam, A., Friedrich, T., Shaxson, F. and Pretty, J. (2009), "The spread of conservation agriculture: justification, sustainability and uptake", *International Journal of Agricultural Sustainability*, Vol. 7 No. 4, pp. 292-320.

Kassie, M., Heilemraim, T., Marenya, P., Jaleta, M. and Erenstein, O. (2015), "Production risks and food security under alternative technology choices in Malawi: application of a multinomial endogenous switching regression", *Journal of Agricultural Economics*, Vol. 66 No. 3, pp. 640-659.

Kassie, M., Jaleta, M., Shiferaw, B., Mmbando, F. and Mekuria, M. (2013), "Adoption of interrelated sustainable agricultural practices in smallholder systems: evidence from rural Tanzania", *Technological Forecasting and Social Change*, Vol. 80 No. 3, pp. 525-540.

Kassie, M., Zikhali, P., Pender, J. and Köhlin, G. (2010), "The economics of sustainable land management practices in the Ethiopian highlands", *Journal of Agricultural Economics*, Vol. 61 No. 3, pp. 605-627.

Khatri-Chhetri, A., Aryal, J.P., Sapkota, T.B. and Khurana, R. (2016), "Economic benefits of climate-smart agricultural practices to smallholder farmers in the Indo-Gangetic plains of India", *Current Science*, Vol. 110 No. 7, pp. 1244-1249.

Lal, R. (1997), "Residue management, conservation tillage and soil restoration for mitigating greenhouse effect by CO_2-enrichment", *Soil and Tillage Research*, Vol. 43 Nos 1/2, pp. 81-107.

Lal, R. (2004), "Soil carbon sequestration impacts on global climate change and food security", *Science*, Vol. 304 No. 5677, pp. 1623-1627.

Liebman, M. and Dyck, E. (1993), "Crop rotation and intercropping strategies for weed management", *Ecological Applications*, Vol. 3 No. 1, pp. 92-122.

Lin, B.B. (2011), "Resilience in agriculture through crop diversification: adaptive management for environmental change", *BioScience*, Vol. 61 No. 3, pp. 183-193.

Mundlak, Y. (1978), "On the pooling of time series and cross section data", *Econometrica*, Vol. 46 No. 1, pp. 69-85.

Neufeldt, H., Jahn, M., Campbell, B.M., Beddington, J.R., DeClerck, F., De Pinto, A., Gulledge, J., Hellin, J., Herrero, M., Jarvis, A., LeZaks, D., Meinke, H., Rosenstock, T., Scholes, M., Scholes, R., Vermeulen, S., Wollenberg, E. and Zougmoré, R. (2013), "Beyond climate-smart agriculture: toward safe operating spaces for global food systems", *Agriculture & Food Security*, Vol. 2 No. 1, p. 12.

Pender, J.L. and Kerr, J.M. (1998), "Determinants of farmers' indigenous soil and water conservation investments in semi-arid India", *Agricultural Economics*, Vol. 19 Nos 1/2, pp. 113-125.

Pingali, P.L. (2012), "Green revolution: impacts, limits, and the path ahead", *Proceedings of the National Academy of Sciences*, Vol. 109 No. 31, pp. 12302-12308.

Rao, N.H. (2007), "A framework for implementing information and communication technologies in agricultural development in India", *Technological Forecasting and Social Change*, Vol. 74 No. 4, pp. 491-518.

Sapkota, T.B., Jat, M.L., Aryal, J.P., Jat, R.K. and Khatri-Chhetri, A. (2015), "Climate change adaptation, greenhouse gas mitigation and economic profitability of conservation agriculture: Some examples from cereal systems of Indo-Gangetic plains", *Journal of Integrative Agriculture*, Vol. 14 No. 8, pp. 1524-1533.

Sapkota, T.B., Majumdar, K., Jat, M.L., Kumar, A., Bishnoi, D.K., McDonald, A.J. and Pampolino, M. (2014), "Precision nutrient management in conservation agriculture based wheat production of northwest India: Profitability, nutrient use efficiency and environmental footprint", *Field Crops Research*, Vol. 155, pp. 233-244.

Sarkar, R.K., Reddy, J.N., Sharma, S. and Ismail, A.M. (2006), "Physiological basis of submergence tolerance in rice and implications for crop improvement", *Current Science*, Vol. 91 No. 7, pp. 899-906.

Sehgal, V.K., Singh, M.R., Chaudhary, A., Jain, N. and Pathak, H. (2013), *Vulnerability of Agriculture to Climate Change: District Level Assessment in the Indo-Gangetic Plains*, Indian Agricultural Research Institute, New Delhi, ISBN: 978-81-88708-97-0.

Shiferaw, B., Aryal, J.P. and Mittal, S. (2012), "Cereal system diversification for sustainable productivity growth, climate risk management and adaptation in South Asia", *3rd International Agronomy Congress*, New Delhi.

Singh, R.B. (2000), "Environmental consequences of agricultural development: a case study from the green revolution state of Haryana, India", *Agriculture, Ecosystems & Environment*, Vol. 82 Nos 1/3, pp. 97-103.

Teklewold, H., Kassie, M. and Shiferaw, B. (2013), "Adoption of multiple sustainable agricultural practices in rural Ethiopia", *Journal of Agricultural Economics*, Vol. 64 No. 3, pp. 597-623.

Wollni, M., Lee, D.R. and Thies, J.E. (2010), "Conservation agriculture, organic marketing, and collective action in the Honduran hillsides", *Agricultural Economics*, Vol. 41 Nos 3/4, pp. 373-384.

Wu, J. and Babcock, B.A. (1998), "The choice of tillage, rotation, and soil testing practices: economic and environmental implications", *American Journal of Agricultural Economics*, Vol. 80 No. 3, pp. 494-511.

Yamano, T., Rajendran, S. and Malabayabas, M.L. (2015), "Farmers' self-perception toward agricultural technology adoption: evidence on adoption of submergence-tolerant rice in Eastern India", *Journal of Social and Economic Development*, Vol. 17 No. 2, pp. 260-274.

Yu, L., Hurley, T., Kliebenstein, J. and Orazem, P. (2012), "A test for complementarities among multiple technologies that avoids the curse of dimensionality", *Economics Letters*, Vol. 116 No. 3, pp. 354-357.

About the authors

Jeetendra Prakash Aryal is a Climate Economist with the Socioeconomics Program in the International Maize and Wheat Improvement Center (CIMMYT). He joined CIMMYT from May 2012 and has been working in Climate Change, Agriculture and Food Security (CCAFS) in South Asia. His current works focus in identifying coping and adaptation strategies of farmers and the poor to manage future climate change. In addition, his work analyses the economic incentives and benefits to farmers from adoption of conservation agriculture for adaptation to and mitigating climate change in south Asia. Before joining CIMMYT, he worked as a researcher in the Norwegian University of Life Sciences, Norway. He had also served Tribhuvan University, Nepal as a lecturer of Economics. His main areas of interest are land productivity, land reforms, technology adoption, conservation agriculture, analysis of risk coping strategies of the farm households and climate change. Aryal obtained his MA in Mathematical Economics from Tribhuvan University, Nepal and MSc and PhD in Development and Resource Economics from the Norwegian University of Life Sciences, Norway.

M.L. Jat is a Senior Cropping Systems Agronomist (Principal Scientist) with CIMMYT's Sustainable Intensification Program and currently responsible for coordinating CIMMYT-CCAFS research in South Asia. Before joining CIMMYT, he served Indian Council of Agricultural Research (ICAR) for 12 years as Systems Agronomist and Conservation Agriculture Specialist. His significant research contributions in the fields of conservation agriculture, precision farming and climate change in major production systems of South Asia are well-documented in nearly 200 publications. A well-recognized expert in the field of conservation agriculture (CA) and climate smart agriculture (CSA), M.L. is a recipient of large number of awards, honors and fellowships from national and international organizations and societies.

Tek B. Sapkota is Climate Change and Agricultural System Scientist working in Sustainable Intensification Program in the International Maize and wheat improvement Centre (CIMMYT). He obtained PhD in agro-biosciences (agriculture, environment and landscape) with specialization in nutrient dynamics in agro-ecosystem. His research interest includes analysis of cropping systems from food security climate change nexus. He is involved in studying management consequences on nutrient dynamics in agro-ecosystem and their effect on food security, poverty reduction, climate change adaptation and mitigation.

Arun Khatri-Chhetri is an agricultural economist at CCAFS-BISA, India. Prior to joining CCAFS, he has worked as a Research Consultant for Conservation International and International Food Policy Research Institute (IFPRI) Washington D.C. USA. He also worked as Research Associate in Davis College of Agriculture, Natural Resources and Design, West Virginia University, USA. He spent one year as a visiting researcher at the School of International Development and Cooperation, Hiroshima

University, Japan. Dr Arun has over eight years of research and development experiences on socio-economic and institutional aspects of natural resource management and conservation, agricultural production and food security, and vulnerability to climate change and adaptations. His key interests include agricultural and water resource management, adaptation options for climate change, sustainable use and protection of natural resources and ecosystems for human well-being, and market-based approach of natural resource management and conservation. Arun completed his PhD in Agriculture and Natural Resource Economics from West Virginia University, USA and has a Master degree in Agricultural Economics from Tribhuvan University, Nepal.

Menale Kassie is a Development Economist working at the International Centre of Insect Physiology and Ecology (ICIPE). Before joining ICIPE in 2015, he had been with the International Maize and Wheat Improvement Center (CIMMYT) for about five years. In CIMMYT, he coordinated the socioeconomic components of the program "Sustainable intensification of maize-legume cropping systems for food security in eastern and southern Africa (SIMLESA)" supported by the Australian Government through the Australian Center for International Agricultural Research (ACIAR). His research focuses on adoption and impact of crop and natural resource management technologies on rural household welfare, using advanced cross-section-panel econometrics and mathematical programing models.

Dr Dil Bahadur Rahut is currently working in International Maize and Wheat Improvement Center (CIMMYT) as Science Program Manager, Socioeconomics Program, CIMMYT. He holds a PhD in Development Economics from the University of Bonn, MA International Political Economy from Tsukuba University, Japan and MBA (Financial Management) from India. Before joining CIMMYT, Dil worked at South Asian University, Bank of Bhutan Limited, Indian Council for Research on International Economic Relations (ICRIER), WorldFish Centre, Centre for Development Research (ZEF) and Royal Monetary Authority of Bhutan. Applied economic research including households' decision-making under climate change and resource allocation are the research areas of Dil Bahadur Rahut. Dil Bahadur Rahut is the corresponding author and can be contacted at: dilbhutan@gmail.com

Sofina Maharjan is an Economist working in the International Maize and Wheat Improvement Centre (CIMMYT). She is a consultant for the project entitled "Sustainable and Resilient Farming System Intensification (SRFSI)". She has worked in several projects related to climate change such as Climate Change, Agriculture and Food Security (CCAFS) and Adoption and Impact of Laser Land Leveling in India, She received MA in economics from the South Asian University, India.

2

Measuring the level of corporate commitment regarding climate change strategies

Iordanis Eleftheriadis and Evgenia Anagnostopoulou
Department of Business Administration, University of Macedonia, Thessaloniki, Greece

Abstract

Purpose – This study aims to examine the various climate change practices adopted by firms and develop a set of corporate indexes that measure the level of climate change corporate commitment, climate change risk management integration and climate change strategies adoption. Moreover, this study examines the relationship between the aforementioned indexes. The authors claim that there is a positive relationship between the adoption of climate change strategies, corporate commitment and risk management integration. The aforementioned indexes have been used to assess the largest companies in the oil and gas sectors.

Design/methodology/approach – To assess this study's sample companies, a content analysis of their carbon disclosure project (CDP) reports for the years 2012-2015 was conducted. Finally, weights were assigned to the content analysis data based on the results of a survey regarding the difficulty of implementing each climate change practice included in the respective index. The survey sample included climate change experts who are either currently employed in companies that are included in the Financial Times Global 500 (FT 500) list, or work as external partners with these companies.

Findings – The present study results highlight the need for developing elaborate corporate indexes, as the various climate change practices have different degrees of difficulty regarding their implementation. Additionally, a general trend in adopting climate change strategies is observed, especially in the field of carbon reduction strategies, which mainly involve the implementation of low carbon technologies. Finally, a positive and significant relationship was found between carbon reduction targets, risk management integration and climate change strategies.

Practical implications – Although international research has extensively examined the importance of managers' perceptions on environmental issues as an enabling factor in developing environmental strategies, according to the results of our survey, corporations must go beyond top management commitment towards climate change to be able to successfully implement climate change strategies. Incorporation of climate change risk management procedures into a company's core business activities as well as the establishment of precise carbon reduction targets can provide the basis on which successful climate change strategies are implemented.

Originality/value – Most studies address the issue of climate change management in terms of environmental or sustainability management. Furthermore, research on climate change and its relationship with business management is mainly theoretical, and climate change corporate performance is measured with aggregate indexes. This study focuses on climate change which is examined from a five-dimensional perspective: top management commitment, carbon reduction targets, risk management integration, carbon reduction and carbon compensation strategies. This allows us to conduct an in-depth analysis of the various climate change practices of firms.

Keywords Risk management, Climate change strategies, Corporate commitment

1. Introduction

Climate change has been widely acknowledged as one of the major sources of risk by the global community. Taking as a reference the current flow of CO_2 emissions, it has been estimated that there is a chance ranging between 77 and 99 per cent that the global temperature will rise more than 2°C in the next 20 years (Stern, 2006). Such an increase in global temperature is most likely to cause rapid changes to current climatic models, affecting directly the natural environment. However, the risks that climate change poses are not confined strictly to the environmental and physical impacts of global temperature rise but they also involve social, economic and financial impacts (Cuevas, 2011).

The world-wide acceptance that climate change, as a result of the rise of CO_2 emissions concentration in the atmosphere, is attributed to human activities has triggered the adoption of policies both at national and international levels. These climate change policies aim to put a price on carbon emissions, for example, though carbon taxes, the establishment of carbon trading programs such as the European Union Emissions Trading System (ETS) (Chevallier, 2009), the setting of mandatory processes and product standards or the provision of incentives to invest in low carbon technologies (Bebbington and Larrinaga-Gonzalez, 2008). Moreover, institutional investors, banks, accounting firms, governmental agencies, NGOs and consumers have begun to demand information disclosure regarding the corporate climate change practices of firms (Maria Gonzalez-Gonzalez and Zamora Ramírez, 2016). Subsequently, there is a pressing need for businesses to develop appropriate climate change strategies for the risks posed by the projected climate change policies.

The aim of this study is to examine the various climate change practices adopted by firms and to develop a set of corporate climate change indexes that measure the level of climate change corporate commitment, the level of climate change risk management integration (RMI) into business activities and the level of adoption of corporate climate change strategies. Moreover, we examine the relationship between the aforementioned indexes. We claim that there is a positive relationship between the adoption of climate change strategies, corporate commitment and climate change RMI. Specifically, we develop two corporate indexes regarding corporate commitment: top management commitment (TMC) and carbon reduction targets (CRTs). Regarding climate change strategy, we distinguish between carbon reduction strategies (CRS) and carbon compensation strategies (CCS). Finally, we develop one index for climate change RMI. We use the aforementioned indexes to assess the largest companies in the oil and gas sectors for the years 2012-2015.

To assess our sample companies, we conducted a content analysis of their carbon disclosure project (CDP) reports for the years 2012-2015. Finally, we assigned weights to the content analysis data based on the results of a survey regarding the difficulty of implementing each climate change practice included in the respective index. The survey sample included climate change experts who are either currently employed in companies that are included in the Financial Times Global 500 (FT 500) list, or work as external partners with these companies.

The structure of this paper is as follows: in the next chapter, we conduct a review of previous research related to climate change strategies. Along with the relevant literature review, we develop our climate change indexes and present our research hypotheses. Then, we describe our scoring methodology, the development of the data weights and our regression model. Finally, we present the results of our survey, provide a discussion of our findings, outline the limitations of our study and make propositions for future research.

2. Conceptual framework

Regarding climate change, it is only recently that firms have begun to treat it as more than a corporate social responsibility issue. According to Porter and Reinhardt (2007), business leaders need to carefully examine the cost of emissions to their business as well as a firm's vulnerability to physical, economic and social impacts of climate change. Studying a firm's value chain and assessing its exposure to climate change can help them develop strategies that will not only reduce current and impeding climate change risks but also reveal business opportunities and enhance corporate performance. There are a number of different strategies that a firm can adopt and those depend on both the strategic choices of top management and its available resources (Lee, 2012; Christmann, 2000). Some firms may choose to make incremental changes regarding their business activities, while others may choose to make radical changes on their business model.

In one of the earliest papers on climate change strategy, Dunn (2002) briefly describes the technological, economic and policy implications of climate change on firms. Regarding the technological dimension, special attention is given to the use of alternative fuel sources, such as natural gas and renewable energy sources, as well as the development of efficient combined heat and power technologies. Also, the use of market-based instruments, such as carbon trading, is suggested as a way of lowering the cost of reducing CO_2 emissions. Weinhofer and Hoffmann (2010) divide climate change strategies into three different groups:

(1) *CO_2 compensation strategies*, which involve actions taken by firms to balance their carbon emissions by participating in emissions trading systems (ETSs), buying certified emission reductions credit units (CERs) or investing in emissions reduction projects.

(2) *CO_2 reduction strategies*, which focus on the improvement of CO_2 emitting production processes or on the design of new products, whose production emits fewer emissions.

(3) *Carbon independence strategies*, which are based on the design of processes and products that are carbon-free or which radically reduce carbon emissions.

Finally, Kolk and Pinkse (2008) also distinguish between innovation strategies and compensation strategies. They describe innovation strategies as process improvements that reduce energy consumption, such as the installation of energy-efficient technologies for carbon-intensive industries or the development of processes focused on the reduction of CO_2 emissions through the supply chain. Regarding compensation strategies, in the same line with Weinhofer and Hoffmann (2010), these include the internal transfer of carbon emissions through emissions trading and participation in carbon offset projects.

In this study, we develop two corporate indexes that measure the level of adoption of climate change strategies: the CCS index and the CRS index. Regarding the CCS index, in line with the aforementioned research, it incorporates the participation in emissions trading schemes, the creation of project-based carbon credits and the purchase of carbon credits. In relation to the CRS index, particular attention is placed on carbon-friendly technology and carbon reduction business processes. In this study, we are going to examine some indicative climate change technologies based on the research of Cadez and Czerny (2016). The following carbon reduction technologies are examined: natural gas, combined heat and power, renewable energy sources, carbon capture and storage and increased boiler efficiency. Finally, we examine the relationship between the aforementioned strategy indexes, two climate change corporate commitment indexes and one RMI index.

Regarding climate change corporate commitment and risk management, NGOs and research organizations have released several categorization schemes. More specifically, the CDP, the Coalition for Environmental Responsible Economies (Ceres), the Global Reporting Initiative (GRI), the Climate Disclosure Standards Board (CDSB) and RobecoSam classify managerial climate change strategies into three, broadly defined, categories: *Top Management Commitment*, which involves corporate commitment to climate change performance, top management responsibility and executive compensation; *Carbon Reduction Targets*; and *Climate Change Risk Management Integration* into core business strategy.

2.1 Relationship between corporate commitment and climate change strategy

Regarding climate change governance, high-level managers are generally considered as those who are in a position to make changes in an organization, to choose the business environment in which the firm operates and to engrave its course in the long term. In Linnenluecke *et al.*'s (2015) study on executives' perception on the need of developing climate change adaptation strategies, the authors find that engagement with climate change science and the perceived degree of the firm's vulnerability are positively related to the choice of developing climate change strategies. Therefore, examining the level of TMC in climate change is important in understanding the way climate change strategies are implemented within organizations.

The association between TMC and environmental or ethical commitment has been examined by international research (Boiral *et al.*, 2012, Aragón-Correa *et al.*, 2004; Weaver *et al.*, 1999; Henriques and Sadorsky, 1999). Results show that TMC in environmental issues is positively associated with increased corporate environmental commitment. Regarding climate change commitment, Jeswani *et al.* (2008) separate corporate climate change strategies into two categories: operational activities for energy efficiency and GHG reduction and management activities. Along those lines, TMC towards GHG reductions, the adoption of GHG reduction targets (both absolute emission reduction targets and business intensity targets) and the development of a corporate GHG inventory are examined as enabling factors for developing successful climate change strategies.

Moreover, Lee (2012) also proposes a set of carbon management activities. These include setting GHG reduction targets and developing a new market and new business activities. Moreover, organizational involvement, which is defined as top management involvement in climate change initiatives and encouragement of employees to undertake climate change initiatives, is also examined (Lee, 2012). Finally, the role of the vision, quality and skills of managers in the adoption of climate change strategies is also highlighted in the research of Boiral (2006). In this paper, we examine two aspects of corporate climate change commitment. The first one involves the commitment of top management towards climate change performance and the second is the adoption of CRTs. Based on the above, we develop the following hypotheses:

H1. There is a positive association between climate change strategy and top management commitment towards climate change.

H2. There is a positive association between climate change strategy and carbon reduction targets.

2.2 Relationship between risk management and climate change strategy

Climate change is characterized by a high degree of uncertainty and by the lack of sufficient prior data related to extreme weather events which can lead to "massive discontinuous changes" (Winn *et al.*, 2011). For that reason, the statistical methods based on historical data, the traditional probabilistic models used by insurance companies and the risk management analysis tools that exist today may prove to be insufficient in integrating quantitative and qualitative climate change risks into corporate operational activities (Winn *et al.*, 2011). This highlights the need to develop different risk management procedures that incorporate the principles of climate change risk management, climate change adaptation and resilience strategies into one integrated framework.

Perceived firm vulnerability to climate change physical risks is closely related to the perceived impacts of climate change, which include both the direct impacts on a firm's business operations and the indirect impacts on a firm's market and business environment. Moreover, it is affected by a firm's past experience with climate change risks, by the lessons learnt from previous extreme weather events and by its ability to quantify and assess the financial implications of extreme weather events (Pinkse and Gasbarro, 2016; Linnenluecke et al., 2011; Linnenluecke et al., 2008).

All of the above affect a firm's response to the physical risks of climate change. According to Pinkse and Gasbarro (2016), firms respond to climate change risk by taking either "routine measures" or "non-routine measures". Routine measures include developing risk monitoring and assessment procedures, taking technical measures to endure impacts, using financial instruments for risk sharing, developing emergency and restorations plans, etc. On the other hand, non-routine measures involve actions such as assessing the vulnerability of geographic sites of business activities, conducting product portfolio diversification by investing in alternative products or business procedures and driving cooperation within the industry to reduce climate change exposure.

Finally, implementing strategies, which enhance the adaptation and resilience of firms, greatly improves their ability to protect themselves to the adverse implications of climate change. However, for firms to successfully adjust to extreme weather events, they have to link both adaptation and resilience processes into one single action framework (Linnenluecke et al., 2011). Building on the above, we form the following research hypothesis:

H3.   There is a positive association between climate change strategy and climate change risk management integration.

3. Methodology

The aim of this study is to examine the various climate change strategies adopted by firms and to develop a set of corporate climate change indexes that measure the level of corporate commitment regarding climate change, the level of climate change RMI into business activities and the level of adoption of corporate climate change strategies. Based on extensive literature review, we developed five indexes related to climate change corporate practices:

(1) *Top management commitment* (TMC), which involves top management engagement and accountability.

(2) *Climate change risk management integration* (RMI), which examines how risk management processes are implemented by firms.

(3) *Carbon reduction targets* (CRT), long-term and short-term absolute reduction targets and intensity reduction target.

(4) *Carbon reduction strategies* (CRS), which include the development of carbon-efficient technologies and the implementation of business processes that reduce CO_2 emissions.

(5) *Carbon compensation strategies* (CCS), which involves carbon trading and CO_2 emission offsetting projects.

In total, 23 items were developed under these factors. The specific corporate practices related to each factor, and the related research, are presented in Table I. To measure the

Table I. Climate change indexes and related research

Climate change index	Corporate practices	Related research	Item code
Top management commitment (TMC)	The Board of Directors (or a committee of the Board) is accountable for climate change performance	CDP (2015), GRI (2015), RobecoSam (2015), Ceres (2014), CDSB (2012), Boiral (2006) , Hoffman (2005), Renukappa *et al.* (2013)	ITEM_01
	Company management has clear responsibilities for achieving climate change goals	CDP (2015); GRI (2015), RobecoSam (2015); Ceres (2014), CDSB (2012), Hoffman (2005), Renukappa *et al.* (2013)	ITEM_02
	Executive compensation (monetary) is linked to climate change performance	CDP (2015), GRI (2015), RobecoSam (2015); Ceres (2014), Renukappa *et al.* (2013)	ITEM_03
Risk management integration (RMI)	Climate change risk management processes are integrated into core business risk management processes	CDP (2015), GRI (2015), CDSB (2012), Boiral (2006), Lash and Wellington (2007), Hoffman (2005)	ITEM_04
	Climate change risks and opportunities are identified at asset level	CDP (2015), GRI (2015), CDSB (2012); Hoffmann (2006)	ITEM_05
	Climate change risks and opportunities are identified at company level	CDP (2015), GRI (2015), CDSB (2012); Hoffman (2005)	ITEM_06
	Company has processes that allow the prioritization of risks and opportunities related to climate change	CDP (2015), GRI (2015), CDSB (2012)	ITEM_07
Carbon reduction targets (CRT)	Company has short-term absolute[a] CO_2 emission reduction targets	CDP (2015), CDSB (2012), Lee (2012), Jeswani *et al.* (2008), Hoffman (2005), Dunn (2002)	ITEM_08
	Company has short-term CO_2 emission intensity[b] reduction targets	CDP (2015), CDSB (2012), Lee (2012), Jeswani *et al.* (2008), Hoffman (2005), Dunn (2002)	ITEM_09
	Company has long-term absolute CO_2 emission reduction targets	CDP (2015), CDSB (2012), Lee (2012), Jeswani *et al.* (2008), Hoffman (2005), Dunn (2002)	ITEM_10
	Company has long-term CO_2 emission intensity reduction targets	CDP (2015), CDSB (2012), Lee (2012), Jeswani *et al.* (2008), Hoffman (2005), Dunn (2002)	ITEM_11
Carbon Reduction Strategies (CRS)	Fossil fuel switching, from coal to natural gas	Kotchen and Mansur (2016), Lamb *et al.* (2015), Cadez and Czerny (2016), IEA (2015), Dunn (2002)	ITEM_12
	Increased boiler efficiency, by implementing the best available technology	Cadez and Czerny (2016), Qu *et al.* (2014), Li *et al.* (2014), Namioka *et al.* (2012)	ITEM_13
	Usage of Combined Heat and Power Technology (cogeneration)	Gibson *et al.* (2016), Cadez and Czerny (2016), Lund and Mathiesen (2015), Klaassen and Patel (2013), Mago and Smith (2012)	ITEM_14
	Energy source switching, from fossil fuels to renewable energy sources	Cadez and Czerny (2016), IEA (2015), da Graça Carvalho (2012), Boiral (2006), Hoffmann (2006), Neuhoff (2005), Dunn (2002)	ITEM_15

(continued)

Table I.

Climate change index	Corporate practices	Related research	Item code
	Capture and storage of CO2	Cadez and Czerny (2016), IEA (2015), Boot-Handford *et al.* (2014), Gerbelová *et al.* (2013), Scott *et al.* (2013), Gunter *et al.* (2009)	ITEM_16
	Replacement of carbon-based products by non-carbon based products[c]	Cadez and Czerny (2016), Weinhofer and Hoffmann (2010), Jeswani *et al.* (2008)	ITEM_17
	Implementation of end-use energy efficiency processes[d]	Cadez and Czerny (2016), Jeswani *et al.* (2008), Boiral (2006), Hoffman (2005)	ITEM_18
	Optimization of current business processes in order to reduce CO2 emissions	Cadez and Czerny (2016), Weinhofer and Hoffmann (2010), Jeswani *et al.* (2008), Boiral (2006), Hoffman (2005), Kolk and Pinkse (2008)	ITEM_19
	Control of non-CO2 gas emissions (e.g. CH_4, H_2O)	Omara *et al.* (2016), Subramanian *et al.* (2015), Cadez and Czerny (2016), Brantley *et al.* (2014), Dunn (2002)	ITEM_20
Carbon compensation strategies (CCS)	Participating in emissions trading schemes	CDP (2015), Cadez and Czerny (2016), Lee (2012), Jeswani *et al.* (2008), Boiral (2006); Hoffman (2005), Kolk and Pinkse (2008), Dunn (2002)	ITEM_21
	Creating project-based carbon credits	CDP (2015), Cadez and Czerny (2016), Lee (2012), Jeswani *et al.* (2008), Boiral (2006); Hoffman (2005), Kolk and Pinkse (2008), Dunn (2002)	ITEM_22
	Purchasing carbon credits	CDP (2015), Cadez and Czerny (2016), Lee (2012), Jeswani *et al.* (2008), Boiral (2006); Hoffman (2005), Kolk and Pinkse (2008), Dunn (2002)	ITEM_23

Notes: [a]Absolut emission reduction refers to CO_2 reduction in absolute numbers, regardless of business activities; [b]Intensity ratios express CO_2 impact per unit of physical activity or unit of economic output; [c]e.g.replace plastic products with wooden ones; [d]e.g. turning down heating and cooling during non-working hours, reduce non-necessary travel, etc

corporate indexes developed in this study, we conducted a content analysis of the CDP reports of the largest companies in the oil and gas sectors for the years 2012-2015. Finally, we assigned weights to the content analysis data based on the results of a survey regarding the difficulty of implementing each climate change practice included in the respective index.

3.1 Scoring methodology
The scoring methodology used in this study was based on a two-step procedure described later in the text. First, we conducted a content analysis of the CDP reports for each company by using the items included in the climate change indexes described above. "Content analysis has been defined as a systematic, replicable technique for compressing many words of text into fewer content categories based on explicit rules of coding" (Berelson, 1952; GAO, US General Accounting Office, 1996; Krippendorff, 1980; Weber, 1990). It is a method which "enables researchers to sift through large volumes of data with relative ease in a systematic fashion" (GAO, US General Accounting Office, 1996). Moreover, it allows us to "discover and

describe the focus of individual, group, institutional, or social attention (Weber, 1990)." (Stemler, 2001). The method of content analysis has been used by many authors in studying corporate environmental disclosure reports (Freedman and Jaggi, 2005; Brammer and Pavelin, 2006; Prado-Lorenzo et al., 2009; Gallego-Alvarez, 2010; Martínez-Ferrero et al., 2015). We will also apply this methodology to analyze information provided in the CDP reports of oil and gas companies for the years 2012-2015. Each climate change index item takes the value of either 1 if the relevant information is reported or 0 if it is not.

After the content analysis was conducted, we assigned weights to the content analysis data based on the results of a survey regarding the difficulty of implementing each climate change practice included in the respective index. The reason we decided to weigh the content analysis data is because not all climate change practices have the same level of difficulty in their implementation. In Table II, we provide an example, which explains the reason why we decided to weigh the results of the content analysis. We measure the level of corporate commitment for CRS. As we can see, each company has different management practices. If we were to use simple content analysis, then corporate level of commitment for both companies would be the same, which in our example, is equal to 5. However, when we apply weights to the content analysis data, we observe that the level of corporate commitment for Company A (21) is higher than that of Company B (16).

3.2 Development of data weights

We assigned weights to the content analysis data based on the results of a survey regarding the difficulty of implementing each climate change practice described in Table I. The questionnaire was sent to experts in climate change corporate strategies. Respondents were asked to rate each climate change practice under a seven-point Likert scale (1 = Very Easy ... 7 = Very Difficult), to indicate the difficulty level of implementing the respective corporate practice. The target sample involved climate change experts, who are either currently employed in companies that are included in the Financial Times Global 500 (FT 500) list or work as external partners with these companies. We selected climate change experts who are employed by large multinational companies or work with them because these companies have the organizational and financial capabilities to implement a large variety of climate change corporate practices. Therefore, we can assume that climate change experts employed, in this type of organizations, will have extensive experience on climate change corporate practices and can provide a more reliable assessment of the difficulty level regarding their implementation. Target respondents were identified via LinkedIn according to their expertise. Expertise was determined according to the target respondent's skills. To be included in the target sample, each respondent had to have the following skills in their profile: 1st skill: climate change; 2nd skill (at least one the following): strategy, management, strategic management, business strategy. Finally, for the skills mentioned above, target respondents had to have at least 30 endorsements from other members of the LinkedIn Network to be included in the target population of the research.

The questionnaires were sent via the LinkedIn InMail messaging service. In total, 332 questionnaires were sent and 188 complete questionnaires were received, representing a response rate of 56.62 per cent. To increase response rate, three reminders were sent to each target respondent. The first was after one week from the initial email posting, the second after two weeks from the initial posting and the third after four weeks from the initial email posting. The collection of questionnaires began on October 15, 2015, and completed on the March 23, 2016. The descriptive statistics of the final sample are presented in Table III.

Table II. Example of scoring methodology for the carbon reduction strategies index

Carbon reduction strategies	Fossil fuel switching, from coal to natural gas	Increased boiler efficiency	Use of combined heat and power technology	Energy source switching, from fossil fuels to renewable energy sources	Capture and storage of CO_2	Replacement of carbon-based products by non-carbon-based products	Implementation of end-use energy efficiency processes	Optimization of current business processes in order to reduce CO_2 emissions	Control of non-CO_2 gas emissions	Total unweighted score	Total weighted score
Median Values	4	3	4	4	7	5	2	3	4		
Company A	0	1	1	1	1	0	0	1	0	5	21
Company B	1	1	0	1	0	0	1	1	0	5	16

Table III. Questionnaire sample descriptive statistics

Sample characteristics	Response (%)	Response count
Gender		
Male	83.80	158
Female	16.20	30
Total	100.00	188
Age		
18-29	2.70	5
30-44	29.70	56
45-59	54.10	102
60+	13.50	25
Total	100.00	188
Level of education		
High school degree	0.00	0
Bachelor degree	18.90	36
Graduate degree	81.10	152
Total	100.00	188
Job description		
Engineering	2.70	5
Finance/Accounting	0.00	0
Human Resources	0.00	0
Management	48.60	91
Manufacturing	2.80	5
Project Management	5.40	10
Research	2.80	5
Risk Management	2.70	5
Sales/Marketing	0.00	0
Strategy/Planning	18.00	34
Other	17.00	32
Total	100.00	188
Job level		
Executive/C-Level	13.60	26
Senior Management	31.20	59
Middle Management	28.40	53
Intermediate	5.40	10
Entry Level	0.00	0
Other	21.40	40
Total	100.00	188
Industry sector		
Aerospace & defence	1.60	3
Automobiles & parts	3.19	6
Banks	11.17	21
Beverages	1.60	3
Chemicals	5.32	10
Construction & materials	6.38	12
Electricity	10.64	20
Financial services	6.38	12
Telecommunications	3.72	7
Food & drug retailers	1.06	2
Food producers	1.60	3

(*continued*)

Table III.

Sample characteristics	Response (%)	Response count
Gas, water & utilities	6.38	12
General industrials	3.19	6
General retailers	2.13	4
Household goods & home construction	0.53	1
Industrial engineering	1.06	2
Metals & mining	4.26	8
Industrial transportation	1.06	2
Life insurance	0.53	1
Media	1.06	2
Nonlife insurance	2.66	5
Oil & gas producers	10.11	19
Oil equipment & services	4.26	8
Personal goods	2.13	4
Pharmaceuticals	1.06	2
Real estate investment	2.13	4
Software & computer services	0.53	1
Support services	0.53	1
Technology hardware & equipment	0.00	0
Tobacco	0.53	1
Other	3.19	6
Total	100.00	188
Experience in climate change related issues		
Less than 1 year	0.00	0
At least 1 year but less than 3 years	5.40	10
At least 3 years but less than 5 years	5.40	10
At least 5 years but less than 10 years	16.20	30
10 years or more	73.00	137
Total	100.00	188

3.3 Regression model

After we conducted a weighted content analysis of the information in disclosed CDP reports of the sample firms, we developed a dependency model using the CCS and CCR indexes as dependent variables and the TMC, CRT and RMI indexes as independent variables.

$$CRS = f \, (TMC, \, CRT, \, RMI) \tag{1}$$

$$CCS = f \, (TMC, \, CRT, \, RMI) \tag{2}$$

The aforementioned model was empirically estimated using the following equation:

$$CRS_i = \beta_0 + \beta_1 TMC_i + \beta_2 CRT_i + \beta_3 RMI_i + \varepsilon$$

$$CCS_i = \beta_0 + \beta_1 TMC_i + \beta_2 CRT_i + \beta_3 RMI_i + \varepsilon$$

where CRS_i is the carbon reduction strategy index; CCS_i is the carbon compensation strategy index; TMC_i is the top management commitment index; CRT_i is carbon reduction target index; and RMI_i is the risk management integration index.

The aforementioned model was empirically tested using linear regression analysis and estimated using the ordinary least squares methodology.

4. Results
4.1 Survey results' descriptive data
The descriptive statistics of the climate change practices are presented in Table IV. As we can observe, the skewness and kurtosis values range between ±1 which indicate that data are close to normal distribution. Furthermore, regarding the TMC index, ITEM_03, which corresponds to the executive management compensation being linked to climate change performance targets, has the highest mean among the rest of the items. As far as CRTs is concerned, targets related to absolute CO_2 emission reductions generally receive a higher mean of implementation difficulty than intensity targets. Regarding carbon reduction technologies, *carbon capture and storage* has the highest mean difficulty, which can easily be explained if we take into account the fact that it is the newest carbon reduction technology and that it is considered to be at a demonstration stage. Finally, regarding CCS, participation in ETS and creation of project-based carbon credit have higher mean difficulty values than purchasing carbon credits.

4.2 Weighted content analysis descriptive data
Table V presents the descriptive statistics for the oil and gas companies for the years 2012-2015. According to above descriptive statistics, the adoption level of climate change practices varies to a great extent between companies. TMC and RMI have gradually increased for oil and gas companies during the past four years. Additionally, adoption of

Table IV. Survey results: descriptive statistics

Item code	Mean	Median	Skewness statistic	SE of skewness	Kurtosis	SE of Kurtosis
ITEM_01	4.22	4	−0.16	0.177	−0.732	0.353
ITEM_02	3.99	4	0.246	0.177	−0.947	0.353
ITEM_03	4.99	5	−0.482	0.177	−0.9	0.353
ITEM_04	4.12	4	0.111	0.177	−0.853	0.353
ITEM_05	3.62	3	0.58	0.177	−0.513	0.353
ITEM_06	3.76	4	0.337	0.177	−0.755	0.353
ITEM_07	4.03	4	0.196	0.177	−0.88	0.353
ITEM_08	4.8	5	−0.09	0.177	−0.617	0.353
ITEM_09	4.34	4	0.253	0.177	−0.813	0.353
ITEM_10	4.87	5	−0.471	0.177	−0.817	0.353
ITEM_11	4.47	5	−0.261	0.177	−0.908	0.353
ITEM_12	3.66	4	0.201	0.177	−0.247	0.353
ITEM_13	3.11	3	0.761	0.177	0.074	0.353
ITEM_14	3.64	4	0.227	0.177	−0.642	0.353
ITEM_15	3.76	4	0.071	0.177	−0.973	0.353
ITEM_16	6.37	7	−0.964	0.177	0.212	0.353
ITEM_17	4.64	5	−0.204	0.177	−0.507	0.353
ITEM_18	2.85	2	0.846	0.177	−0.001	0.353
ITEM_19	3.13	3	0.422	0.177	−0.609	0.353
ITEM_20	4.04	4	0.1	0.177	−0.967	0.353
ITEM_21	4.63	5	−0.555	0.177	−0.237	0.353
ITEM_22	4.82	5	−0.599	0.177	0.326	0.353
ITEM_23	3.38	3	0.394	0.177	0.111	0.353

Table V. Descriptive statistics for oil and gas companies

Index	2012				2013				2014				2015			
	Mean	Median	Minimum	Maximum	Mean	Median	Minimum	Maximum	Mean	Median	Minimum	Maximum	Mean	Median	Minimum	Maximum
Board Oversight Climate Change Risk Management	8.66	8	0	13	9.81	11	4	13	9.41	11	4	13	9.97	11	4	13
Integration	11.84	15	0	15	12.06	15	0	15	12.44	15	0	15	12.53	15	0	15
Carbon Reduction Targets	4.53	4	0	14	4.84	4	0	14	3.72	4	0	14	3.75	4	0	10
Carbon Reduction Strategies	9.19	10	0	21	9.91	10	0	21	11.5	13	0	25	12.38	13.5	0	25
Carbon Compensation Strategies	5.31	5	0	13	6.03	5	0	13	6.41	6.5	0	13	4.84	5	0	13

CRTs has slightly decreased in 2014 and 2015 for oil and gas companies, while the rate of CRS has increased. On the other hand, we observe that the adoption of CCS increased until 2014 and decreased in 2015. Summarizing our results indicate a general trend in investing in low carbon technologies and a gradual shift from compensation strategies to CRS.

4.3 Results of empirical analysis

4.3.1 Correlation statistics. Table VI presents the bivariate correlations between the variables. The correlations between the variables are all positive and significant at the 0.01 level. The highest correlations are detected between RMI and both climate change strategy indexes. Furthermore, the CRT index also shows a very high correlation to the CCS.

4.3.2 Regression analysis. To estimate the ordinary least squares regression function, we analysed several statistical assumptions of the regression analysis, such as normality, homoscedasticity, multicollinearity and autocorrelation. Regarding normality, we applied the Kolmogorov–Smirnov test, which showed us that the variables do not exhibit a normal distribution. The lack of normal distribution is owing to the size of our sample and the existence of extreme values in our data. That is because firms will disclose a large amount of information, very little or none at all. Nevertheless, following Gallego-Alvarez (2010) and Lumley *et al.* (2002), we assume that the lack of a normal distribution does not affect the quality of our results. Regarding autocorrelation in the residuals from the regression, we conducted the Durbin-Watson test. The value obtained from the Durbin-Watson test is approximately 2, which reflects the absence of autocorrelation in the residuals. The variance inflation factors (VIFs) and tolerance factors have been used to analyse the absence or presence of multicollinearity. For there to be no multicollinearity problems, the values obtained in tolerance have to be high and the values obtained in the VIFs have to be low. The collinearity statistics presented in Table VII show that our models present tolerances

Table VI. Pearson's correlation

Correlations	TMC	RMI	CRT	CRS	CCS
TMC	1	0.438**	0.292**	0.276**	0.376**
RMI	0.438**	1	0.381**	0.503**	0.615**
CRT	0.292**	0.381**	1	0.415**	0.507**
CRS	0.276**	0.503**	0.415**	1	0.344**
CCS	0.376**	0.615**	0.507**	0.344**	1

Note: **Correlation is significant at the 0.01 level (two-tailed)

Table VII. Regression statistics

	Model 1	Collinearity statistics			Model 2	Collinearity statistics	
	CRS	Tolerance	VIF	CCS		Tolerance	VIF
TMC	0.029	0.790	1.266	0.084		0.790	1.266
CRT	0.257**	0.738	1.355	0.307**		0.738	1.355
RMI	0.392**	0.836	1.197	0.461**		0.836	1.197
R Square	0.312			0.471			
F	18.311**			35.923**			
Durbin-Watson	1.842			1.935			

Notes: *$p < 0.05$; **$p < 0.01$

between 0.790 and 0.836, and the VIF between 1.187 and 1.335, indicating the absence of multicollinearity.

Regarding the explanatory power of our models, these have R^2 values of 0.312 and 0.471 for a confidence level of 99 per cent ($p < 0.01$). These values are similar to those obtained in relevant studies. For example, Freedman and Jaggi (2005) obtained a value of 0.310 for the R^2, Prado-Lorenzo *et al.* (2009), 0.383, and Gallego-Alvarez (2010), 0.406. The results from the regression analysis, using the ordinary least squares methodology, are presented in Table VII.

For a confidence level of 99 per cent, CRTs and RMI have a positive and statistically significant effect on CRS and CCS. On the other hand, there is a positive but not significant relationship between TMC and both climate change strategy indexes. These results allow us to accept *H2* and *H3* and to reject *H1*.

Our results highlight the importance of climate change RMI into core business activities as an enabling factor in implementing corporate climate change strategies. In contrast to the results of relevant research, the effect of TMC on climate change strategies is not significant, while the adoption of CRTs is significantly related to both strategy indexes. Subsequently, setting CRTs and incorporating climate change risk management practices in core business practices contributes much more to the successful development of climate change strategies than TMC. These findings are particularly interesting because they stress on the need for business to go beyond the traditional corporate governance approach to environmental issues towards the establishment of a more concrete basis for implementing climate change strategies.

5. Conclusion

Climate change has been globally acknowledged as a major source of physical, economic and social risks. Companies are expected increased costs in their production processes and their supply chains, which will gradually affect their profitability. Moreover, the increased social and economic risks have also caught the attention of various stakeholders, such as institutional investors, banks, accounting firms, governmental agencies, NGOs and consumers who have been demanding information regarding corporate climate change practices. This study aimed to examine the various climate change practices adopted by firms and factors that influence the adoption of climate change corporate strategies. Subsequently, we developed two corporate indexes regarding climate change corporate commitment: *top management commitment* and *carbon reduction targets*. Regarding climate change strategy, we distinguished between *carbon reduction strategies* and *carbon compensation strategies*. Finally, we developed one index for climate change, *risk management integration*. We used the above indexes to assess the largest companies in the oil and gas sectors for the years 2012-2015.

Very few studies have attempted to address the issue of climate change management separately from the general concepts of environmental or sustainability management. Furthermore, the majority of research on climate change and its relationship with business management is theoretical and attempts to measure climate change performance that is based on the use of single aggregate indexes. Thus, the novelty of this study is that it focuses on only climate change and second that it does not use aggregate indexes to measure climate change corporate practices. Instead, we examine five different dimensions of climate change management which allow us to conduct an in-depth analysis of the various climate change practices of firms.

Our results show that there is a significantly positive relationship between carbon compensation and CRS, climate change RMI and CRTs. On the other hand, the relationship between the two carbon strategy indexes and TMC is deemed insignificant based on our results. This has practical implications for business, as it highlights the necessity of a coherent corporate climate change basis to enhance the success of both CRS and CCS. Although international research has extensively examined the importance of managers' perceptions on environmental issues as an enabling factor in developing environmental strategies, according to the results of our survey, corporations must go beyond TMC towards climate change to be able to successfully implement climate change strategies. Incorporation of climate change risk management procedures into a company's core business activities and the establishment of precise CRTs can provide the basis on which successful climate change strategies are implemented.

Our research has a number of limitations. First, to weigh our corporate climate change indexes, we used survey data based on the opinions of climate change experts. However, although our sample population was formed by experts who work in large multinational corporations, the data collected were based on the opinion of experts who work in multiple business sectors. In conducting our survey, we had to compromise with experts working in different sectors and not specific to the oil and gas sector, owing to the fact that our target population was very limited. We believe that in the future, we will be able to locate more experts of each respective sector to have more precise responses regarding the difficulty level of implementing climate change practices.

Future research could examine other economic sectors, and do comparative research between carbon-intensive and non-carbon-intensive sectors. Moreover, our indexes, used to measure corporate commitment regarding climate change strategies, could be examined in relation to corporate performance data. For example, researchers could use the indexes developed in the present study as dependent or independent variables in regression models to examined hypotheses related to the effect of climate change management in the financial performance of firms. This would also contribute to the general discussion regarding the win–win hypothesis of implementing both climate change strategies and achieving enhanced financial performance.

References

Aragón-Correa, J.A., Matías-Reche, F. and Senise-Barrio, M.E. (2004), "Managerial discretion and corporate commitment to the natural environment", *Journal of Business Research*, Vol. 57 No. 9, pp. 964-975.

Bebbington, J. and Larrinaga-Gonzalez, C. (2008), "Carbon trading: accounting and reporting issues", *European Accounting Review*, Vol. 17 No. 4, pp. 697-717.

Berelson, B. (1952), *Content Analysis in Communication Research*, Free Press, Glencoe, IL.

Boiral, O., Henri, J.F. and Talbot, D. (2012), "Modeling the impacts of corporate commitment on climate change", *Business Strategy and the Environment*, Vol. 21 No. 8, pp. 495-516.

Boiral, O. (2006), "Global warming: should companies adopt a proactive strategy?", *Long Range Planning*, Vol. 39 No. 3, pp. 315-330.

Boot-Handford, M.E., Abanades, J.C., Anthony, E.J., Blunt, M.J., Brandani, S., Mac Dowell, N., Fernández, J.R. and Ferrari, M.-C. (2014), "Carbon capture and storage update", *Energy & Environmental Science*, Vol. 7 No. 1, pp. 130-189.

Brammer, S.J. and Pavelin, S. (2006), "Corporate reputation and social performance: the importance of fit", *Journal of Management Studies*, Vol. 43 No. 3, pp. 435-455.

Brantley, H.L., Thoma, E.D., Squier, W.C., Guven, B.B. and Lyon, D. (2014), "Assessment of methane emissions from oil and gas production pads using mobile measurements", *Environmental Science & Technology*, Vol. 48 No. 24, pp. 14508-14515.

Cadez, S. and Czerny, A. (2016), "Climate change mitigation strategies in carbon-intensive firms", *Journal of Cleaner Production*, Vol. 112, pp. 4132-4143.

CDP (2015), "Climate change strategies and risk management – the perspective of companies and investors", *CDP CEE 100 Climate Change Report 2014*.

CDSB (2012), "Climate change reporting framework: advancing and aligning disclosure of climate change-related information in mainstream reports", *The Climate Disclosure Standards Board*.

Ceres (2014), *Gaining Ground: Corporate Progress on the Ceres Roadmap for Sustainability*, Ceres.

Chevallier, J. (2009), "Emissions trading: what makes it work?", *International Journal of Climate Change Strategies and Management*, Vol. 1 No. 4, pp. 400-406.

Christmann, P. (2000), "Effects of 'best practices' of environmental management on cost advantage: the role of complementary assets", *Academy of Management Journal*, Vol. 43 No. 4, pp. 663-680.

Cuevas, S.C. (2011), "Climate change, vulnerability, and risk linkages", *International Journal of Climate Change Strategies and Management*, Vol. 3 No. 1, pp. 29-60.

Da Graça Carvalho, M. (2012), "EU energy and climate change strategy", *Energy*, Vol. 40 No. 1, pp. 19-22.

Dunn, S. (2002), "Down to business on climate change: an overview of corporate strategies", *Greener Management International*, Vol. 2002 No. 39, pp. 27-41.

Freedman, M. and Jaggi, B. (2005), "Global warming, commitment to the Kyoto protocol, and accounting disclosures by the largest global public firms from polluting industries", *The International Journal of Accounting*, Vol. 40 No. 3, pp. 215-232.

Gallego-Alvarez, I. (2010), "Indicators for sustainable development: relationship between indicators related to climate change and explanatory factors", *Sustainable Development*, Vol. 20 No. 4, pp. 276-292.

GAO, US General Accounting Office (1996), *Content Analysis: A Methodology for Structuring and Analyzing Written Material*, GAO/PEMD-10.3.1, Washington, DC.

Gerbelová, H., Versteeg, P., Ioakimidis, C.S. and Ferrão, P. (2013), "The effect of retrofitting Portuguese fossil fuel power plants with CCS", *Applied Energy*, Vol. 101, pp. 280-287.

Gibson, C.A., Meybodi, M.A. and Behnia, M. (2016), "A methodology to compare combined heat and power systems operating under emissions reduction policies considering biomass co-fired, coal- and natural gas-fuelled steam turbines", *Energy Efficiency*, Vol. 9 No. 6, pp. 1271-1297.

GRI (2015), "G4 sustainability reporting guidelines", *The Global Reporting Initiative*.

Gunter, W.D., Bachu, S., Buschkuehle, M., Michael, K., Ordorica-Garcia, G. and Hauck, T. (2009), "Reduction of GHG emissions by geological storage of CO_2", *International Journal of Climate Change Strategies and Management*, Vol. 1 No. 2, pp. 160-178.

Henriques, I. and Sadorsky, P. (1999), "The relationship between environmental commitment and managerial perceptions of stakeholder importance", *Academy of Management*, Vol. 42 No. 1, pp. 187-199.

Hoffman, A.J. (2005), "Climate change strategy: the business logic behind voluntary greenhouse gas reductions", *California Management Review*, Vol. 47 No. 3, pp. 21-46.

Hoffmann, A.J. (2006), *Getting Ahead of the Curve: Corporate Strategies That Address Climate Change*, The Pew Center on Global Climate Change, Arlington, VR.

IEA (2015), *World Energy Outlook Special Report 2015: Energy and Climate Change*, International Energy Agency.

Jeswani, H.K., Wehrmeyer, W. and Mulugetta, Y. (2008), "How warm is the corporate response to climate change? Evidence from Pakistan and the UK", *Business Strategy and the Environment*, Vol. 17 No. 1, pp. 46-60.

Klaassen, R.E. and Patel, M.K. (2013), "District heating in the Netherlands today: a techno-economic assessment for NGCC-CHP (natural gas combined cycle combined heat and power)", *Energy*, Vol. 54, pp. 63-73.

Kolk, A. and Pinkse, J. (2008), "A perspective on multinational enterprises and climate change: learning from 'an inconvenient truth' and quest", *Journal of International Business Studies*, Vol. 39 No. 8, pp. 1359-1378.

Kotchen, M.J. and Mansur, E.T. (2016), "Correspondence: reassessing the contribution of natural gas to US CO_2 emission reductions since 2007", *Nature Communications*, Vol. 7, pp. 10648.

Krippendorff, K. (1980), *Content Analysis: An Introduction to Its Methodology*, Sage, Newbury Park, CA.

Lamb, B.K., Edburg, S.L., Ferrara, T.W., Howard, T., Harrison, M.R., Kolb, C.E., Townsend-Small, A. and Dyck, W. (2015), "Direct measurements show decreasing methane emissions from natural gas local distribution systems in the United States", *Environmental science & Technology*, Vol. 49 No. 8, pp. 5161-5169.

Lash, J. and Wellington, F. (2007), "Competitive advantage on a warming planet", *Harvard Business Review*, pp. 95-102.

Lee, S.Y. (2012), "Corporate carbon strategies in responding to climate change", *Business Strategy and the Environment*, Vol. 21 No. 1, pp. 33-48.

Li, Y.Y., Hou, S.S. and Sheu, W.J. (2014), "Investigation on boiler efficiency and pollutant emissions of water/heavy oil emulsions using edge-tone resonant homogenizer", *Fuel*, Vol. 119, pp. 240-251.

Linnenluecke, M., Griffiths, A. and Winn, M.I. (2008), "Organizational adaptation and resilience to extreme weather events", paper presented at the Annual Meeting of the Academy of Management, Anaheim, CA, and recipient of the Carolyn Dexter Best International Paper Award.

Linnenluecke, M.K., Griffiths, A. and Mumby, P.J. (2015), "Executives' engagement with climate science and perceived need for business adaptation to climate change", *Climatic Change*, Vol. 131 No. 2, pp. 321-333.

Linnenluecke, M.K., Stathakis, A. and Griffiths, A. (2011), "Firm relocation as adaptive response to climate change and weather extremes", *Global Environmental Change*, Vol. 21 No. 1, pp. 123-133.

Lumley, T., Diehr, P., Emerson, P.S. and Chen, L. (2002), "The importance of the normality assumption in large public health data sets", *Annual Review of Public Health*, Vol. 23 No. 1, pp. 151-169.

Lund, R. and Mathiesen, B.V. (2015), "Large combined heat and power plants in sustainable energy systems", *Applied Energy*, Vol. 142, pp. 389-395.

Mago, P.J. and Smith, A.D. (2012), "Evaluation of the potential emissions reductions from the use of CHP systems in different commercial buildings", *Building and Environment*, Vol. 53, pp. 74-82.

Maria Gonzalez-Gonzalez, J. and Zamora Ramírez, C. (2016), "Voluntary carbon disclosure by Spanish companies: an empirical analysis", *International Journal of Climate Change Strategies and Management*, Vol. 8 No. 1, pp. 57-79.

Martínez-Ferrero, J., Garcia-Sanchez, I.M. and Cuadrado-Ballesteros, B. (2015), "Effect of financial reporting quality on sustainability information disclosure", *Corporate Social Responsibility and Environmental Management*, Vol. 22 No. 1, pp. 45-64, doi: 10.1002/csr.1330.

Namioka, T., Yoshikawa, K., Takeshita, M. and Fujiwara, K. (2012), "Commercial-scale demonstration of pollutant emission reduction and energy saving for industrial boilers by employing water/oil emulsified fuel", *Applied Energy*, Vol. 93, pp. 517-522.

Neuhoff, K. (2005), "Large-scale deployment of renewables for electricity generation", *Oxford Review of Economic Policy*, Vol. 21 No. 1, pp. 88-110.

Omara, M., Sullivan, M.R., Li, X., Subramanian, R., Robinson, A.L. and Presto, A.A. (2016), "Methane emissions from conventional and unconventional natural gas production sites in the marcellus shale basin", *Environmental Science & Technology*, Vol. 50 No. 4, pp. 2099-2107.

Pinkse, J. and Gasbarro, F. (2016), "Managing physical impacts of climate change an attentional perspective on corporate adaptation", *Business & Society*, doi: https://doi.org/10.1177/0007650316648688.

Porter, M.E. and Reinhardt, F.L. (2007), "A strategic approach to climate", *The Harvard Business Review*, Vol. 85 No. 10, pp. 22-26.

Prado-Lorenzo, J.-M., Rodríguez-Domínguez, L., Gallego-Álvarez, I. and García-Sánchez, I.-M. (2009), "Factors influencing the disclosure of greenhouse gas emissions in firms world-wide", *Management Decision*, Vol. 47 No. 7, pp. 1133-1157.

Qu, M., Abdelaziz, O. and Yin, H. (2014), "New configurations of a heat recovery absorption heat pump integrated with a natural gas boiler for boiler efficiency improvement", *Energy Conversion and Management*, Vol. 87, pp. 175-184.

Renukappa, S., Akintoye, A., Egbu, C. and Goulding, J. (2013), "Carbon emission reduction strategies in the UK industrial sectors: an empirical study", *International Journal of Climate Change Strategies and Management*, Vol. 5 No. 3, pp. 304-323.

RobecoSam (2015), "RobecoSAM's corporate sustainability assessment methodology RobecoSAM Ag", available at: www.robecosam.com

Scott, V., Gilfillan, S., Markusson, N., Chalmers, H. and Haszeldine, R.S. (2013), "Last chance for carbon capture and storage", *Nature Climate Change*, Vol. 3 No. 2, pp. 105-111.

Stemler, S. (2001), "An overview of content analysis", *Practical Assessment, Research & Evaluation*, Vol. 7 No. 17, available at: http://PAREonline.net/getvn.asp?v=7&n=17 (accessed 10 March 2016)

Stern, N. (2006), *The Economics of Climate Change: The Stern Review*, Cambridge University Press, Cambridge, MA.

Subramanian, R., Williams, L.L., Vaughn, T.L., Zimmerle, D., Roscioli, J.R., Herndon, S.C. and Sullivan, M.R. (2015), "Methane emissions from natural gas compressor stations in the transmission and storage sector: measurements and comparisons with the EPA greenhouse gas reporting program protocol", *Environmental Science & Technology*, Vol. 49 No. 5, pp. 3252-3261.

Weaver, G.R., TreviÑo, L.K. and Cochran, P.L. (1999), "Integrated and decoupled corporate social performance: management commitments, external pressures, and corporate ethics practices", *Academy of Management*, Vol. 42 No. 5, pp. 5539-5552.

Weber, R.P. (1990), *Basic Content Analysis*, 2nd ed., Newbury Park, CA.

Weinhofer, G. and Hoffmann, V.H. (2010), "Mitigating climate change-how do corporate strategies differ?", *Business Strategy and the Environment*, Vol. 19 No. 2, pp. 77-89.

Winn, M., Kirchgeorg, M., Griffiths, A., Linnenluecke, M.K. and Günther, E. (2011), "Impacts from climate change on organizations: a conceptual foundation", *Business Strategy and the Environment*, Vol. 20 No. 3, pp. 157-173.

Corresponding author

Evgenia Anagnostopoulou can be contacted at: eanagno@uom.edu.gr

3

Perceptions of adaptation, resilience and climate knowledge in the Pacific

Rory A. Walshe
Department of Geography, King's College London, London, UK and
Institute of Risk and Disaster Reduction (IRDR), University College London,
London, UK

Denis Chang Seng
Intergovernmental Oceanographic Commission, UNESCO, Paris, France

Adam Bumpus
School of Geography, Faculty of Science, University of Melbourne,
Melbourne, Australia, and

Joelle Auffray
Apidae Development Innovations Pty Ltd., Melbourne, Australia

Abstract

Purpose – While the South Pacific is often cited as highly vulnerable to the impacts of climate change, there is comparatively little known about how different groups perceive climate change. Understanding the gaps and differences between risk and perceived risk is a prerequisite to designing effective and sustainable adaptation strategies.

Design/methodology/approach – This research examined three key groups in Samoa, Fiji and Vanuatu: secondary school teachers, media personnel, and rural subsistence livelihood-based communities that live near or in conservation areas. This study deployed a dual methodology of participatory focus groups, paired with a national mobile phone based survey to gauge perceptions of climate change. This was the first time mobile technology had been used to gather perceptual data regarding the environment in the South Pacific.

Findings – The research findings highlighted a number of important differences and similarities in ways that these groups perceive climate change issues, solutions, personal vulnerability and comprehension of science among other factors.

Practical implications – These differences and similarities are neglected in large-scale top-down climate change adaptation strategies and have key implications for the design of disaster risk reduction and climate change adaptation and therefore sustainable development in the region.

Originality/value – The research was innovative in terms of its methods, as well as its distillation of the perceptions of climate change from teachers, media and rural communities.

Keywords Perceptions, Climate change, Adaptation, Vanuatu, Samoa, Fiji

1. Introduction

It is widely predicted that the impacts of climate change in the South Pacific will arrive comparatively early and be severely felt (Barnett and Campbell, 2010; Walsh *et al.*, 2012), and climate change and its negative impacts will present major obstacles and challenges to sustainable development in the region (UNOHRLLS, 2009).

It is increasingly appreciated that perceptions of climate change and environmental risk influence the degree and nature of adaptation actions taken at the community level (Adger *et al.*, 2005; Leiserowitz, 2006; Mortreux and Barnett, 2009). Despite this, there is a comparative lack of research and knowledge on how different areas and groups in the South Pacific perceive climate change (Lata and Nunn, 2012). The Human Development Report (United Nations Development Program, 2011), which compiles data for perceptions of the environment, has no figures or indicators for Samoa, Fiji or Vanuatu. Adapting to climate change without considering the knowledge and capacity that exists at the local level is likely to lead to failure and maladaptation (Mercer *et al.*, 2012). Therefore, in order to develop effective adaptation strategies, it is important to understand how different communities and groups in the South Pacific perceive climate change.

This article presents findings from an innovative UNESCO-funded research project (UNESCO, 2014a), conducted between January and September 2013. The project "Sharing Perceptions of Adaptation, Resilience and Climate Knowledge" (SPARCK) used ground-breaking socio-technological methods to survey three key sectors (secondary school teachers, rural communities and media personnel) in Samoa, Fiji and Vanuatu. The authors were directly involved in this research.

This paper contributes to the literature in two principal ways: first, it provides empirical evidence that local perceptions may strongly differ from national recommendations and plans; second, it illustrates an innovative methodology not used before in adaptation planning. As a result, and in the context of the potential climate impacts associated with anthropogenic warming, we contend that to make adaptation planning and implementation effective, local perceptions need to be engaged more thoroughly and that human–technology combined methodologies may provide an efficient and effective approach.

2. Adaptation responses in the pacific
2.1 Climate impacts and resilience in the Pacific
Projections and models suggest that in the South Pacific temperature will increase (Ruosteenoja *et al.*, 2003), rainfall events will become more intense, but possibly less frequent (Jones *et al.*, 1999), sea levels will rise (Walsh *et al.*, 2012), regional climate systems such as ENSO are likely to change and tropical cyclones are predicted to become more intense but less frequent (IPCC, 2013, p. 126).

As a result, it is often stressed that the small island states of the South Pacific are on the "front lines" of climate change (Barnett and Campbell, 2010; Birkmann *et al.*, 2014; Garschagen *et al.*, 2016; Nurse *et al.*, 2014; UNESCO, 2009). While the gravity of the potential impacts of climate change in the South Pacific is well established, it is necessary to avoid immediately associating "islandness" with vulnerability and "problematizing" small islands (Kelman and Khan, 2013). While many attributes of small islands present issues of vulnerability [such as isolation, insularity and small population size etc., (Méheux *et al.*, 2007)], there are equally inherent attributes of small islands which present resilience, often emerging from the same characteristics that justify their labelling as vulnerable (such as community cohesion concomitant with isolation and small population; Gaillard, 2007; Kelman *et al.*, 2011).

2.2 Adaptation approaches so far

While investment in climate change adaptation (CCA) in the South Pacific is growing, most projects have focused on raising awareness and engagement at the community and household level with adaptation strategies contained in national adaptation plans (Etkin and Ho, 2007; Lata and Nunn, 2012). Such top-down approaches which aim to bring communities in line with national adaptation plans and communication strategies are often ineffective because communities (both rural and urban) are likely to be unreceptive to information, tools and methods which they perceive as alien, and, consequently, issues persist with ineffective communication (Nunn, 2009; Patt and Schroter, 2008). Instead, communicating information about CCA and best practices should be grounded in local perceptions and the appropriate cultural context (Leiserowitz, 2006; Lindell and Perry, 2003). Therefore, there is a pressing need to understand how various groups and communities in the South Pacific perceive climate change, to develop effective adaptation strategies.

2.3 Role of perceptions in adaptation longevity

Responses to climate change are mediated by how individuals and communities perceive both the challenges and solutions of climate change (Mortreux and Barnett, 2009). The complex and diverse factors which contribute to perceptions include social networks and capital, media influence (Farbotko, 2005), personal experience, worldviews and values (Dessai *et al.*, 2003; McLeman and Smit, 2006). An additional influence on perceptions of climate change is the confidence in the community to successfully adopt adaptation strategies and actions (Barnett and Adger, 2003). Therefore, perceptions are value laden, highly subjective and indivisible from the cultural and social context (Kunreuther and Slovic, 1996). In the Marshall Islands, Rudiak-Gould (2013) demonstrated that climate change is interpreted in a holistic manner where a mixture of beliefs (such as loss of "clan magic" and biblical exegesis) and local knowledge is blended with scientific knowledge within a narrative of the compromise and decay of modernity. Conversely, Lazrus (2009) showed that communities in Tuvalu are energetic and persistent in their response to climate change. However other SIDS communities have demonstrated surprising indifference to harmful climate change impacts (Orłowska, 2016). Similarly, in other SIDS cases, perceptions have demonstrated the potential to act as barrier to adaptation (Betzold, 2015). For example, climate change can be perceived as a distant problem (both in time and space), which may not come to pass (Mataki *et al.*, 2006; Tompkins, 2005).

Local spiritual beliefs on SIDS can also influence perceptions of climate change. In Tuvalu, for example, sea level rise has been attributed to scripture, and consequently resulted in inaction (Mortreux and Barnett, 2009; Paton and Fairbairn-Dunlop, 2010) and in the Cook Islands climate impacts are enmeshed with traditional leadership, tourism and emerging Christian eco-theologies (Rubow, 2013).

Neglecting these perceptions, and the disconnect between national (top-down) and community (bottom-up) approaches to CCA in the South Pacific is one reason why the prevailing top-down approach pursued by government has not always been successful in eliciting appropriate and sustained adaptation (Lata and Nunn, 2012; Nunn, 2010). Indeed, education and adaptation policies that impose Western scientific concepts on rural communities, which lack formal or Western education, have unsurprisingly failed (Berkes and Jolly, 2002; Mcnamara, 2013), and may also undermine the traditional mechanisms for coping with change (Heyd and Brooks, 2009). Therefore, the plurality of perceptions and approaches within and between social systems needs to be recognised and reconciled to rise to the challenge of adaptation and create locally appropriate responses that are sustainable over time and space (Barnett, 2010).

3. Case studies and methodology
3.1 Case studies: Samoa, Fiji and Vanuatu
This research investigated climate change perceptions in Samoa, Fiji and Vanuatu (Figure 1). The selected countries allowed a broad cross-section of the South Pacific to be examined, with diverse representation from both Melanesia and Polynesia. Our paper does not focus and provide details here concerning the culture or geography of these three countries. Rather, we examine what they have in common, for instance, regarding their exposure to climate change impacts. This includes, for example, less frequent but more severe tropical cyclones in Vanuatu (VMGHD, 2011), coastal erosion and salt-water intrusion in Fiji (Korovulavula, 2016) and increasing temperatures in Samoa (Ministry of Natural Resources and the Environment –MNRE, 2013), as well as a combination of these associated impacts in each country, which interact and cannot be disentangled from other environmental issues. For example, tropical cyclone Pam in Vanuatu recently (March 2015) showed the high damage and loses associated with climatic stressors, underscoring the importance of understanding not only the physical component of climate change impacts but also their social and cultural aspects.

Additionally, these countries were selected because each presented access to participants, including good mobile network coverage for the mobile survey. The diversity within and between countries in the South Pacific means that the results from this research should not be expected to be representative of the South Pacific as a whole.

The main goal of the research was to investigate the perceptions of those groups responsible and empowered for communicating climate change information (Teachers and Media) to a wider group of people, as well as those receiving information and who are exposed to climate change impacts (communities) in Samoa, Fiji and Vanuatu:

3.1.1 Secondary school teachers. The importance of integrating climate change within education, particularly secondary school (ages 12-16 years), to effect behaviour change in line with adaptation is well established (Bangay and Blum, 2010; Wiese, 2012). Several reports stress the need to integrate climate change within the education systems of the South Pacific, with secondary education suggested to have a "trickle up" effect of knowledge transfer from students to parents and the community at-large (Hartmann *et al.*, 2010; UNESCO, 2006; Vize, 2013). However, there is relatively little known about how teachers themselves perceive climate change, and how this subsequently influences education and teaching. Consequently, this research investigated secondary school teachers, who were selected as respondents by snowball sampling in each country with the assistance of local partners and the relevant ministry of education, particularly to generate a diverse sample of ages, gender and subjects taught. Henceforth, this group is referred to as "teachers".

3.1.2 Media personnel. While a growing body of literature highlights the role of media in communicating climate change (Boykoff and Smith, 2010; Wilson, 2003), there is a dearth of research investigating how the media perceive climate change. The media sector in Samoa, Fiji and Vanuatu is dominated by radio and print media (particularly national newspapers)

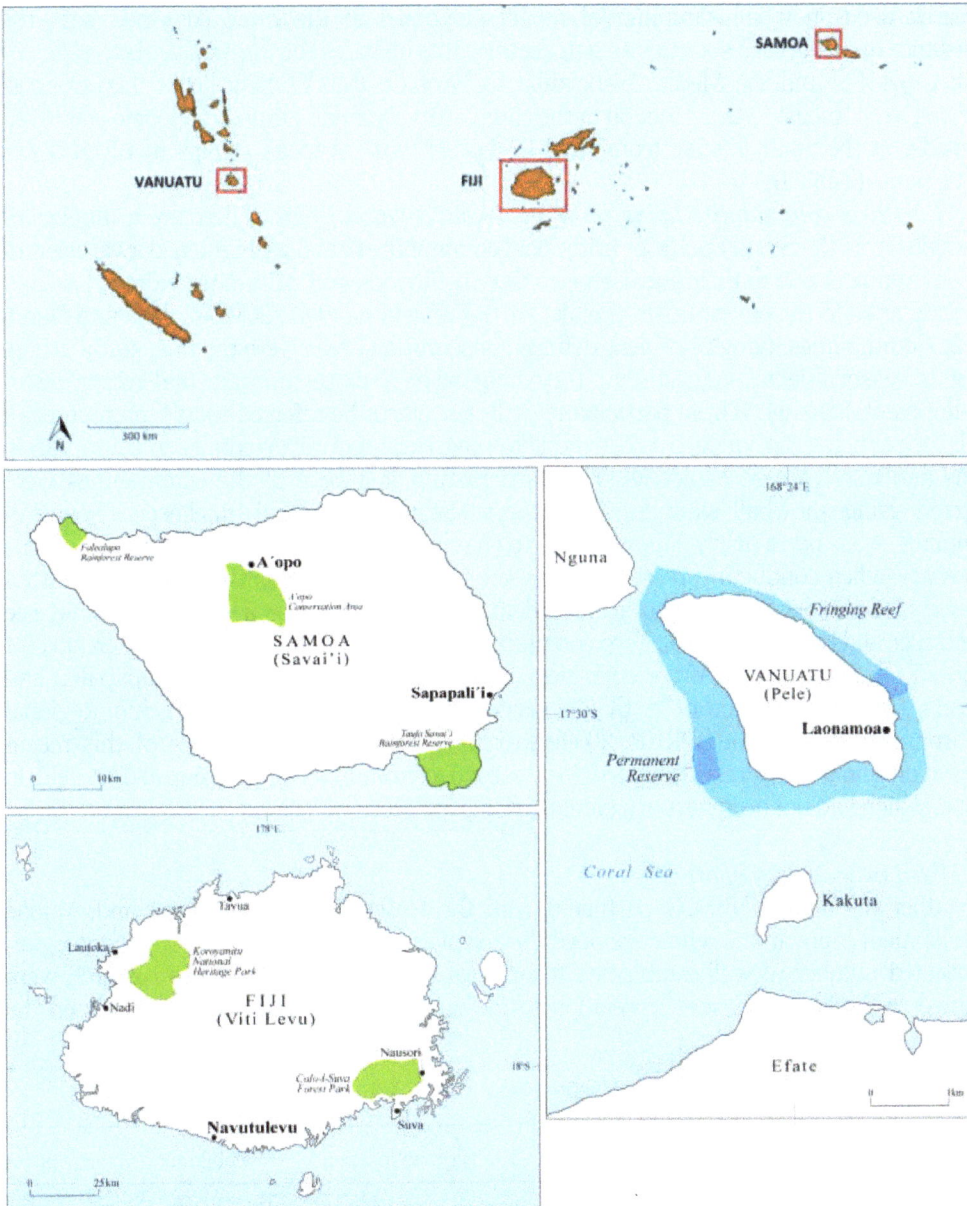

Figure 1. The case study locations in Samoa, Fiji and Vanuatu

(Papoutsaki and Harris, 2008), with a growing television sector. The isolated, disparate and low income nature of many islands in the South Pacific (particularly Vanuatu and Fiji for this study) has resulted in radio playing a central role in community life; however, increasing development and access has resulted in the emergence of digital media, from television to online news, and an increase in newspaper circulation (Papoutsaki and Harris, 2008). In this context, "media personnel" refers to those employed to create (i.e. the journalists and editors) "traditional" media (radio, newspapers, online news and TV). This research used snowball sampling of media personnel in the three countries with the assistance of media civil societies as gatekeepers, this includes the *Journalists Association of Samoa (JAWS)* and the Media Association of Vanuatu (MAV), henceforth this group is referred to as "media". As a function of the comparatively small number of people employed in media in the South Pacific (compared to teachers), these focus groups involved fewer participants (Table I).

3.1.3 Rural communities living in or near conservation areas. There are a number of examples from the South Pacific of traditional communities having heightened awareness of the environment due to their subsistence based livelihoods and culture providing close and intricate links to the environment (Lefale, 2010; Nakashima *et al.*, 2012; Walshe and Nunn, 2012). Communities living in or near defined conservation areas were the final study group, as it is reasonable to suggest that they may have a more intimate understanding of environmental change. These participants will henceforth be referred to as "communities", with the participating villages selected due to their rural nature, proximity to conservation areas and relative ease of access. The participating members of the communities were selected using snowball sampling with the assistance of a local gatekeeper. Snowball sampling was particularly appropriate to navigate the complicated cultural rituals necessary when conducting research in these areas, to gain access to participants with a diversity of perspectives and to achieve demographic parity (for a similar example, see Altschuler and Brownlee, 2015). Two community focus groups were conducted in Samoa (as opposed to just one in the other countries). This was because one village (Sapapali'i) had already been the subject of a CCA intervention (Secretariat of the Pacific Regional Environment Programme-SPREP, 2009); therefore, the potential influence of this recent project on community responses led to a second community focus group being held in A'opo, which had not been part of a climate change project.

3.2 Mixed methodology approach
To gather the data, UNESCO partnered with the University of Melbourne and Apidae Development Innovations who developed the sociological and technological methodologies. A selected number of well-established techniques known in social science research were adapted to the specific issues in relation to the target groups. This section draws on the

Table I. Participants by method, country and group

Groups	Focus group participants	Focus group location	Mobile survey participants
Samoa			
Teachers	33	Apia	55
Media	12	Apia	10
Communities	18/37	A'opo/Sapapali'i	45
Fiji			
Teachers	11	Suva	6
Media	10	Suva	5
Communities	11	Navutulevu	75
Vanuatu			
Teachers	28	Port Vila	25
Media	13	Port Vila	12
Communities	26	Laonamoa	11

methodologies outlined in UNESCO (2014a), which includes full listings of questionnaires and focus group guides.

This research undertook two methods, each of which was applied separately to the three sets of participants (teachers, media and communities) in Samoa, Fiji and Vanuatu.

3.2.1 Interactive focus groups. The participants from each group (selected using snowball sampling and guided by in-country assistance) conducted separate focus groups divided into two activities that prompted discussion of climate change, both of the general experiences of climate change and, in the case of teachers and media, from the perspectives of their profession. To minimise the influence of the positionality or "demand characteristics" between the researchers and participants, throughout both activities, it was stressed that this was about personal experiences, and there were no wrong answers. Furthermore, the activities themselves were designed to be participatory, non-leading and, as far as possible, assisted or facilitated by the local contacts in the local language.

3.2.1.1 Activity 1: problems and solutions identification. Focus group participants were first asked to identify what they perceive as the primary problems of, and solutions to, climate change. The activity was slightly tailored to each group, for example teachers were asked what the problems and solutions were in *teaching* about climate change, media were asked with regards to *communicating* climate change, and communities their *personal experience* of climate change issues and solutions. This was an interactive and participatory activity, with the participants writing their answers on post-it notes to encourage open discussion (Plate 1) which prioritised local opinions and experiences.

Next, a facilitator grouped the issues thematically, putting the problems and solutions post-it notes in related clusters (for example those related to changes in the weather). Participants then voted (using coloured stickers) on these clusters for what they, individually, perceived as the biggest solution, the biggest issue and what they wanted to know more about. This whole first stage was transcribed immediately, with translation if necessary, and the resulting clusters of prioritised issues and solutions were photographed (Plate 2).

3.2.1.2 Activity 2: Q-sort. Q-sort is a collective image sorting method of agreement with a statement, which is then used to prompt discussion (Palmer, 1983; Pitt and Sube, 1979). "Q-methods" were designed to be an objective method for the study of subjective values, opinions and meanings, in conjunction with other methods (Robbins and Krueger, 2000). Its

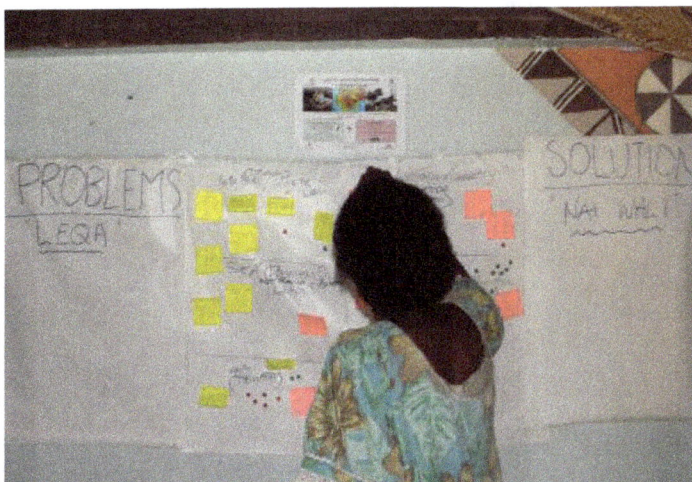

Plate 1. Problems and solutions identification exercise

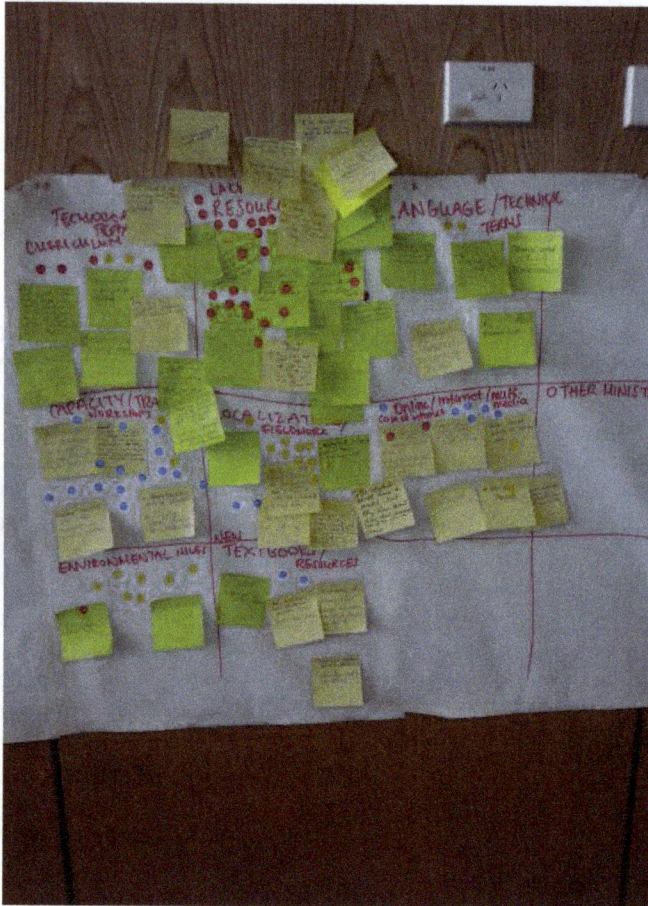

Plate 2. Clusters of prioritised issues and solutions from focus groups

validity and relevance for gauging community perceptions of the environment on small islands is demonstrated by Green (2005). In this case, the participants were divided into small groups and given a stack of images that visually represented various aspects or elements of what could be perceived as climate change (for example, an image of a cyclone). These included a description written on the back, and additional blank cards could be drawn or written on by participants and added. The participants then placed the images on a grid along a continuum according to agreement with a statement, such as "I teach my students about this climate change issue" (teachers only) (Plate 3). This was then discussed as a group and transcribed, as well as photographed. This was used to generate discussion, and not to produce quantitative data, as sometimes is the case in Q-sort applications.

These focus groups were informal and semi-structured, conducted with the assistance of three or more facilitators, which allowed for smaller groups in which issues could be followed up on. Such participatory approaches are particularly useful to understand local perspectives (Chambers and Mayoux, 2003), and have been used effectively in the South Pacific to understand the visualisation of environmental hazards (Cronin *et al.*, 2004). The resulting discussions were recorded in real time, and, in some cases, were held in the local

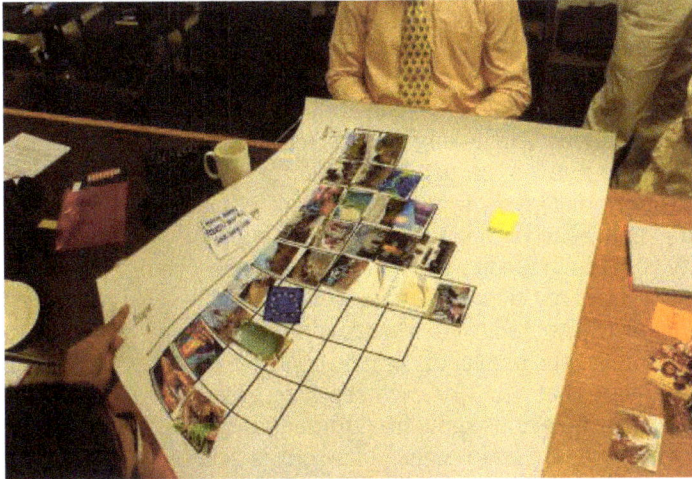

Plate 3. Q-sort activity

language and transcribed immediately with the assistance of a translator. The number of participants varied between each of the focus groups (Table I).

3.2.2 Mobile phone based survey. This research used mobile phones to deliver quantitative and qualitative questions regarding climate change perceptions to specific groups in Samoa, Fiji and Vanuatu. Early applications of mobile and ICT-based methods suggest potential for its use in climate change research in the Pacific (Wadley *et al.*, 2014). Such methods should ideally be twinned with participatory methods to ensure that cultural aspects are accounted for, and are particularly suited as a "probe" approach; one that is used as a means of gathering initial data which leads to further investigation, particularly regarding values (Wadley *et al.*, 2014). As far as the authors are aware, this was the first (and so far only) use of such a method for gathering perceptual data in the South Pacific.

Participants self-selected and self-registered for the mobile survey via simple text message (e.g. "text COM to 894"). This self-registration required a targeted promotion strategy pursuing potential participants from the three groups. This included distributing posters and emails with the help of the relevant government ministries, as well as a social media campaign. The participants of the focus groups were also encouraged to publicise the survey as widely as possible with colleagues in their profession. As opposed to the focus groups (with between 9 and 37 participants in each), the mobile survey was open to as many teachers, media personnel and communities in each country as possible.

Once registered, and depending on the specific group code received with registration, participants were sent 12 text message questions. Aside from a number of key standard questions, each set of questions was tailored to the group (e.g. teachers). Depending on the mobile network, the question messages were either received sequentially via Short Message Service (SMS) or all at once via Unstructured Supplementary Services Data (USSD), to which participants replied via text message. The 12 messages contained concise and carefully worded questions that were sent out in English and in the local language. The survey was then "timed-out" if not completed in a certain time ($\sim$2 weeks). Participation was free of charge, and, as an incentive, participants were entered into a draw to win US$100 in mobile phone credit upon successfully completing the survey.

The first message introduced the survey with a brief explanation that the survey was not a test of knowledge, but instead aimed at gathering opinions. The first three standardised questions were concerned with participant information (gender, age and community district/

teaching subject/media type). This was followed by questions gauging climate change knowledge (e.g. what is the biggest climate change problem?), perceptions of personal risk and vulnerability (e.g. how concerned are you about climate change?) and perceived ability to act (e.g. do you think we can do anything about climate change?). These questions were answered either by replying with a single character linked to multiple-choice options or by replying with a single number on a Likert scale (for example, 1 representing "strongly agree", 5 as "strongly disagree", 6 as "don't know"). The final question was linked to capacity building and asked respondents to type suggestions for what is required to improve capacity (such as in media, "what would be the best thing to help build capacity in the media?"). Once the survey closed, the data were aggregated and results that were clearly erroneous or incomplete were removed. The remaining data sets were verified and validated in an attempt to minimise the impact of demand characteristics by telephoning a number of participants (the square root of N for each separate group in each country; Table I) and orally asking the survey questions again to confirm authenticity and validate their results.

While every effort was made to reduce biases, it is acknowledged that any interaction between "researchers" and "participants" (either by mobile or focus group) involves a positionality which can influence results; consequently, the analysis of data resulting from these methods should be not be treated as representative beyond this sample but instead should be approached as an insight into perceptions. Moreover, as Table I shows, the numbers of participants (total $N{:}443$) for each method, group and location varied considerably. The exact total current number of teachers, media or communities (as defined above) in each country is not available, and as such, it is not possible to provide an indication of statistical representativeness. For example, in 2014, there were 984 secondary school teachers in Samoa (UNESCO, 2014b) compared to this sample of 88 (33 in focus groups and 55 by mobile survey). However, it is important to point out that not all teachers actually teach climate change. As a result, the data should not be extrapolated or generalised to represent any whole group, demographic or region. Instead it offers a "snap-shot" data set, which can be analysed to derive lessons from these participants, which may be suggestive of larger trends.

4. Results
Given the large amount of quantitative and qualitative data generated by this methodology, the results will be detailed by key findings relevant to the scope of this paper and organised by country and group. Therefore, each section does not necessarily detail all elements of both methods (for full results, see UNESCO, 2014a).

4.1 Samoa
4.1.1 Teachers. Samoan teachers agreed that climate change was an important issue; however, they also felt constrained by the lack of training, education and resources and, therefore, felt unable to communicate climate change to their pupils effectively. As a result, teachers identified training as a high priority. Specifically, 65 per cent (N: 36) of teachers in the mobile survey felt that training and workshops for teachers would be the best solution to enable effective teaching about climate change, and that such workshops or training should be in both Samoan and English, locally contextualized and should involve outdoor experiments and fieldtrips to connect scientific theory with local impacts and realities.

It was also expressed that education in schools needed to be streamlined with broader education initiatives outside of school, within families and communities. The focus groups particularly expressed that students should be engaged at an early age and climate change should be fully integrated into the curriculum of appropriate subjects, with key teaching points linked to climate change in others subjects. In the mobile survey, when asked which subject climate change should be included in, 41 per cent (*N*: 22) of teachers sampled felt it should be integrated in all subjects, 33 per cent (*N*: 18) felt it should be part of geography, and only 17 per cent (*N*: 9) believed it should be taught in sciences.

The mobile survey also asked whether teaching about climate change was a priority, and 94 per cent of teachers responded "yes, a lot" or "yes, somewhat". Furthermore, in response to the mobile question "do you feel you understand climate change?", 85 per cent (*N*: 47) of teachers responded "yes, somewhat" or "yes, a lot". Similarly, 98 per cent (*N*:54) also answered "yes" or "yes somewhat" to "are you concerned about climate change". The Samoan teachers that expressed higher concern in climate change also expressed a desire to deal with or respond to climate change on an individual level, indicated by responses to a question regarding how they dealt with climate change., In an attempt to understand perceptions of the ability to act, the mobile survey asked teachers, "Do you think we can do anything to deal with climate change?" While most teachers responded that something could be done, 7 per cent (*N*: 4) of teachers in Samoa responded that the process of climate change cannot be influenced.

4.1.2 Media. Participants in the media focus group felt there was a tension between their professional role and obligation to inform the public about climate change, and the general disinterest of the public in climate change stories, when compared with other topics. Consequently, just over half of the media respondents (55 per cent/*N*: 6) of the mobile survey felt that public education and awareness raising would be the best solution for communicating climate change. This also led to the suggestion that a major issue for communicating climate change in Samoa is the inability to connect it to personal interests in a way that resonates with the general public. Consequently, a perceived solution was the communication of climate change stories using local and personal contexts, as opposed to the "dry" negotiations, science and acronyms. Television and radio were perceived as the two primary media for reaching audiences to communicate climate change issues.

4.1.3 Communities. In both communities (Sapapali'i and A'opo), community members perceived that "extreme" meteorological events and changes in weather (such as increasing temperatures) were the largest local climate change issues. Both Samoan communities also attributed various issues and challenges to climate change that were not, in the scientific sense, connected to anthropogenic climate change, such as economic concerns, and the use of chemicals in agriculture. According to the communities, the solution to these issues is the provision of development assistance by external actors, particularly financial aid.

The mobile survey found that two-thirds of community members sampled believe that climate change was either human induced, or partly human induced and partly natural. In terms of importance, climate change was ranked equally with economic and livelihood concerns. The mobile survey also replicated the perceived solution of external financial assistance, in addition to local education and community based initiatives. The results suggest that Samoan communities largely perceive climate change as real and having a tangible local impact. However, there was a broad range of perceived levels of concern and understanding of climate change, and predictably those who saw themselves as personally threatened were more likely to take action (although high perceived threat levels were not a prerequisite for taking action).

4.2 Fiji

4.2.1 Teachers. Fijian teachers identified a lack of local examples with which to demonstrate impacts of climate change or practical applications for classroom lessons as the primary issue with teaching climate change. Consequently, community-based workshops were seen as a solution, not only to locally contextualise lessons about climate change but also because it was highlighted that students did not necessarily "believe" in climate change, and this disinterest was a barrier to learning. Further, teachers felt that the curriculum did not adequately integrate climate change, and that this should be achieved by including it only in one subject (as opposed to mainstreamed across all subject), and by starting climate change education in primary school.

4.2.2 Media. Fijian media regard the climate change information they receive as too technical and remote, and difficult to translate into Fijian, and as a result, it is rarely translated at all, resulting in a lack of understanding by communities. The media also felt there was an over saturation of both similar and conflicting information from a large number of sources. Media suggested the need for better co-ordination across agencies (including government and NGOs) and media types (radio, print media, television and online) with frequent networking to ensure consistency. The media also proposed connecting the inaccessible technical and global climate change information with local realities and practices by showing local climate change impacts, which would generate content that communities could relate to.

4.2.3 Communities. The community at Navutulevu perceived that climate change is making them increasingly vulnerable to the impact of tropical cyclones, and the weather patterns they are experiencing are changing and becoming increasingly more variable. As a result, the communities were taking steps such as planting mangroves to strengthen shorelines and protect from storm surges. While the community perceived that humans were causing climate change, they also perceived human capacity on a local scale as a solution to climate change impacts and did not believe that climate change was "out of their hands". This partly explains why higher levels of concern correlated with taking action on climate change (as opposed to a reaction of despair leading to inaction); however, men were more likely to take action when compared with women, the reasons for this were not clear from the data. Moreover, the men who felt the most threatened by climate change also thought that actions were available to respond to that threat. However, while technology was seen as a potential solution, the community did not feel they had adequate access or understanding to harness technology as a solution. Consequently, further education was also highlighted as a solution.

4.3 Vanuatu

4.3.1 Teachers. The teachers in the focus group in Port Vila viewed that the curriculum was lacking in terms of its coverage of climate change, and this was a major barrier to teaching climate change, as teachers were uncertain where to place materials within the existing subjects, and resources were not allocated. The teachers sampled were particularly interested in the possibility of, and methods for, integrating traditional knowledge and culture with scientific methods and lessons. Like the teachers in Fiji, those in Vanuatu believed that climate change education should start in primary school; however in contrast to Fiji, they believed that climate change lessons should be mainstreamed into all lessons and subjects, and that teachers should play a role in developing these materials via the sharing of best practices.

The mobile survey uncovered a divide between the priority given to teaching climate change, with half saying they only prioritised it "a little" and the other half saying that they prioritised it "somewhat" or "a lot". Despite this, the majority of teachers sampled felt that they can and must do something about climate change.

4.3.2 Media. The Ni-Vanuatu media perceived themselves as ineffective in communicating climate change; however, they also felt they should take a leading role in raising awareness and believed that as climate change was a relatively new subject, they required training to achieve this. Similar to the other media focus groups, the Ni-Vanuatu media saw a potential solution in making the science and impacts of climate change easier to understand by contextualising and grounding it locally, particularly by using local examples. Including the media in CCA activities from the offset (as opposed to a media release at the conclusion) was also perceived as a strategy that would help the media report more effectively. The mobile survey demonstrated that the media felt the information provided to them was too technical and frequently inaccessible. This also explains why, in response to a subsequent mobile survey question about whether climate change was understood, half the respondents answered "somewhat" with the other half split between "very little" and "very well".

4.3.3 Communities. The Vanuatu community focus group was held at Laonamoa village at the heart of the Nguna-Pele marine reserve (see www.marineprotectedarea.com.vu), north of Efate Island. As a result, it is not surprising that the community prioritised the protection of natural resources as a key issue, particularly for food security and livelihoods. The community also repeatedly highlighted the capacity of traditional knowledge as a component of their livelihoods and its application in responding to challenges such as climate change (among others). However, this capacity of traditional knowledge is perceived as being underutilised and neglected by external stakeholders, particularly those who conduct "climate change training". The community also pointed to the unprecedented nature of changes in the climate and environment as challenging the boundaries of traditional knowledge (for example in regard to seasonal timings of agricultural cycles) and therefore that climate change might potentially erode traditional knowledge. While the mobile survey showed that most community members felt concerned about climate change, it also uncovered a split in priorities, with half the respondents prioritising education and the economy as the most pressing issue. Regardless, the majority of respondents felt threatened by climate change and most also reported already having taken steps to adapt.

As mentioned above, these data can neither be treated as representative of these groups or nations as a whole nor should they be summarised in unity. This sample clearly shows considerable differences in the identification and prioritisation of climate change problems and solutions. For example, 9 per cent (*N*: 10) of mobile survey respondents in Samoa perceived tsunamis as a climate change problem and 6 per cent (*N*: 7) perceived earthquakes as such. In Fiji, the figures were 1 per cent (*N*: 1) and 3 per cent (*N*: 2), respectively. Conversely, in Vanuatu, tsunamis and earthquakes were not seen as climate change problem.

5. Discussion

While there is a general agreement that climate change is a human induced problem and a threat to society, it is clear that there is considerable inter-group and country diversity in perceptions of climate change. This diversity is problematic to integrate into top-down policy-led interventions, particularly those that are multi-national in scale. The diversity of perceptions shows that countries and subdivisions within them should not be viewed collectively or as in unity. The diverse perceptions demonstrated by this research also included a number of conflicting and contradictory perceptions within groups, and complete agreement on any single discussion within the focus groups was rare. This lack of agreement regarding issues and solutions is often smoothed over, as communities are uncritically approached in adaptation programming (Cannon, 2008); consequently, this research suggests that communities should not be portrayed in overly romantic terms which neglect to consider the social divisions within (Forsyth, 2013).

Furthermore, the research has revealed that teachers, media and particularly communities often prioritise other concerns such as poverty, livelihoods and food security, over climate change. This has been remarked upon before in the South Pacific, particularly for subsistence livelihood reliant communities (Iati, 2008; Nunn, 2012). However, this is rarely accounted for in

CCA planning, with programs instead didactically passing down directives and "crowding out" issues which are perceived as greater local concerns (Baldacchino and Kelman, 2014). These divergent priorities illustrate the importance of empowerment, and that for communities in the South Pacific to make informed decisions requires access to appropriate resources (Betzold, 2015). Simply put, CCA should be done *by* communities, not *on* communities (Barnett, 2008, p. 45). It also adds to the assertion that broader climate change thinking fails to acknowledge the diverse experience of climate risks, or the contested nature of many proposed solutions; therefore, discussions about climate action should be based on this diversity of perceived risks and solutions (Forsyth, 2014).

This research also illustrates the artificial nature of the division between DRR and CCA interventions, given that climate change is increasingly being used as a catch all phrase for assigning blame for a range of disasters, including tsunamis (a trend noted by Mercer, 2010; Kelman *et al.*, 2015).

Moreover, the majority of the participants studied in this research agree that something *can* be done at a local scale in response to the impacts of climate change. This capacity to respond and adapt, as well as the historical context of South Pacific island societies successfully overcoming past changes, is neglected in the prevalent discourse about vulnerable islands, in which islanders are passive and "doomed" (Farbotko, 2010). The results of this research suggest that to prepare and implement suitable strategies (either for climate change or other initiatives), the specific socio-cultural context and the key challenges to addressing adaptation and resilience within this context should be further investigated and better understood. Finally, this research offers a number of lessons in terms of its innovative methodology, and particularly, the potential of mobile technology to deliver survey data. There are several clear advantages to this method, given the often disparate and isolated nature of communities in the South Pacific. Fiji for example has over 100 inhabited islands spread over three million square kilometres, and mobile technology has proliferated rapidly and widely in the Pacific (Duffield *et al.*, 2008). While internet-based surveys were considered (and may be more appropriate in the future), the penetration offered by mobile phones is unparalleled, particularly among the older demographic and in more remote areas which might not yet have access to smartphones or internet-enabled devices. There are also clearly limitations and prerequisites necessary to consider with this approach, several of which were demonstrated in this first application. This includes the importance of extensive publicity to ensure adequate participant registration, considering the multiple possible interpretations of such short survey questions between and within groups, the issues with language translations and methodological issues such as the risk of participants contributing random answers or demand characteristics. However, these issues are surmountable with careful analysis and verification of the data and, particularly, when twinned with other methods for means of triangulation. Therefore, this "probe" application suggests that with careful design and support, mobile technology can be used, not only for human-centred climate research in the Pacific region but also to improve the wider efficiency and support of climate change programmes (Wadley *et al.*, 2014).

6. Conclusion
Regardless of the mitigation actions that result from the 2015 UNFCCC policy processes and the Paris Agreement, significant and sustained adaptation will be required in the South Pacific. To design effective adaptation strategies, it has been repeatedly demonstrated that the perceptions of the communities at risk must be considered, otherwise adaptation is likely to fail. This study shows empirical evidence that there is a dissonance in the diverse way teachers, the media and communities perceive and relate to climate change in the South

Pacific. This encompasses issues, solutions, personal vulnerability and comprehension of science among other factors (UNESCO, 2014a). However, there is also common ground in perceptions that climate change is due to human activities and is already having tangible effects. The groups studied largely perceive that something can be done about climate change, and perhaps most importantly, the majority also felt that actions, lessons and communication about climate change and adaptation should be locally grounded and contextualised, including the use of traditional knowledge where possible.

These findings, particularly the diversity of perceptions, are at odds with the prevalent top-down adaptation approaches and education policies, which do not adequately account for these differences and similarities. Regional education programs have been adopted which directly recommend empowering communities in the South Pacific by the use of locally relevant and culturally appropriate climate change education (UNESCO, 2006) and, more recently, the importance of integrating climate change into the curriculum for both DRR and CCA (SPC, 2012). However, there remains a considerable gap between these statements and their actual implementation (Walid, 2017). At the heart of these issues is the struggle to reconcile the mass produced and "one-size-fits-all" adaptation strategies, with the diverse, local and cultural understanding and experience of climate change (Glantz, 1988; Hulme, 2016). Therefore, further research should be aimed at better understanding the influence of culture and cultural practices on these perceptions.

The mixed methodology approach deployed by this research combined technological innovation (particularly the application of mobile technology) with human/social approaches to provide empirical evidence of local perceptions. This also suggests its potential as an efficient and effective approach to design and support adaptation strategies.

7. Acknowledgements

The authors wish to thank Sue Vize, Kevin Petrini, Anne Meldau, Leauga Tamasoalii Saivaise, Tapu Tuailmafu, Cathy Nunn, Christopher Bartlett, Filomena Nelson, Ali Hobbs and Carol Young. Institutions which provided support included: In Samoa, the Ministry of Education, Sports and Culture (MESC), the Global Climate Change Alliance (GCCA) of the University of the South Pacific (USP) and the Journalists Association of Samoa (JAWS). In Fiji, the Department of Environment, the Ministry of Education, Natural Heritage, Culture and Arts, the Ministry of Information and National Archives and Vodafone Fiji. In Vanuatu, the Ministry of Education, Vanuatu National Advisory Board on Climate Change and Disaster Risk Reduction, the Vanuatu Cultural Centre and the Media Association of Vanuatu (MAV). Thanks are also owed to all colleagues at UNESCO Apia Office for the Pacific States, and Apidae Development Innovations Pty Ltd. The national maps for this publication were kindly produced and provided by the University of Melbourne cartographer. Finally to all the participants, vinaka vaka levu, tank yu tumas and fa'afetai tele lava.

Research funding: The funding for this project was provided by the United Nations Educational Scientific and Cultural Organisation (UNESCO).

References

Adger, W.N., Arnell, N.W. and Tompkins, E.L. (2005), "Successful adaptation to climate change across scales", *Global Environmental Change*, Vol. 15 No. 2, pp. 77-86.

Altschuler, B. and Brownlee, M. (2015), "Perceptions of climate change on the Island of Providencia", *Local Environment*, Vol. 21 No. 5, pp. 615-635.

Baldacchino, G. and Kelman, I. (2014), "Critiquing the pursuit of Island sustainability", *Shima*, Vol. 8 No. 2, pp. 1-21.

Bangay, C. and Blum, N. (2010), "Education responses to climate change and quality: two parts of the same agenda?", *International Journal of Educational Development*, Vol. 30 No. 4, pp. 359-368.

Barnett, J. (2008), "The effect of aid on capacity to adapt to climate change: insights from Niue", *Political Science*, Vol. 60 No. 1, pp. 31-45.

Barnett, J. (2010), "Adapting to climate change: three key challenges for research and policy-an editorial essay", *Wiley Interdisciplinary Reviews: Climate Change*, Vol. 1 No. 3, pp. 314-317.

Barnett, J. and Adger, W.N. (2003), "Climate dangers and atoll countries", *Climatic Change, Kluwer Academic Publishers*, Vol. 61 No. 3, pp. 321-337.

Barnett, J. and Campbell, J. (2010), *Climate Change and Small Island States: Power, Knowledge, and the South Pacific*, Earthscan, London.

Berkes, F. and Jolly, D. (2002), "Adapting to climate change: social-ecological resilience in a Canadian western arctic community", *Conservation Ecology*, Vol. 5 No. 2, p. 18.

Betzold, C. (2015), "Adapting to climate change in small Island developing states", *Climatic Change*, Vol. 133 No. 3, pp. 481-489.

Birkmann, J., Garschagen, M., Mucke, P. and Schauder, A. (2014), *World Risk Report 2014*, UNU-EHS (Ed.), Bonn.

Boykoff, M.T. and Smith, J. (2010), "Media presentations of climate change", in Lever-Tracy, Constance (Ed.), *Routledge Handbook of Climate Change and Society*, Routlede International Handbooks, Abingdon, pp. 210-218.

Cannon, T. (2008), "Reducing people's vulnerability to natural hazards: communities and resilience", Wider Research Paper, p. 19.

Chambers, R. and Mayoux, L. (2003), "Reversing the paradigm: quantification and participatory methods", EDIAIS Conference on New Directions in Impact Assessment for Development: Methods and Pracice, *Manchester*, pp. 1-24.

Cronin, S.J., Gaylord, D.R., Charley, D., Alloway, B.V., Wallez, S. and Esau, J.W. (2004), "Participatory methods of incorporating scientific with traditional knowledge for volcanic hazard management on Ambae Island, Vanuatu", *Bulletin of Volcanology*, Vol. 66 No. 7, pp. 652-668.

Dessai, S., Adger, W.N., Hulme, M., Koehler, J., Turnpenny, J., Warren, R. and Köhler, J. (2003), "Defining and experiencing dangerous climate change defining and experiencing dangerous climate change", *Climatic Change*, Vol. 64 Nos 1/2, pp. 11-25.

Duffield, D.L., Watson, A.H.A. and Hayes, D.M. (2008), "Media and communication capacities in the Pacific region", *Ejournalist*, Vol. 8 No. 1, pp. 20-34.

Etkin, D. and Ho, E. (2007), "Climate change: perceptions and discourses of risk", *Journal of Risk Research*, Vol. 10 No. 5, pp. 623-641.

Farbotko, C. (2005), "Tuvalu and climate change: constructions of environmental displacement in the Sydney morning Herald", *Geografiska Annaler: Series B, Human Geography*, Vol. 87 No. 4, pp. 279-293.

Farbotko, C. (2010), "Wishful sinking: disappearing Islands, climate refugees and cosmopolitan experimentation", *Asia Pacific Viewpoint*, Vol. 51 No. 1, pp. 47-60.

Forsyth, T. (2013), "Community-based adaptation: a review of past and future challenges", *Wiley Interdisciplinary Reviews: Climate Change*, Vol. 4 No. 5, pp. 439-446.

Forsyth, T. (2014), "Climate justice is not just ice", *Geoforum, Elsevier Ltd*, Vol. 54, pp. 230-232.

Gaillard, J. (2007), "Resilience of traditional societies in facing natural hazards", *Disaster Prevention and Management: An International Journal*, Vol. 16 No. 4, pp. 522-544.

Garschagen, M., Hagenlocher, M., Comes, M. and Dubbert, M. (2016), *World Risk Report 2016*, UNU-EHS, Bonn.

Glantz, M.H. (1988), *An Essay on the Interactions Between Climate and Society*, National Center for Atmospheric Research, CO.

Green, R. (2005), "Community perceptions of environmental and social change and tourism development on the Island of Koh Samui, Thailand", *Journal of Environmental Psychology*, Vol. 25 No. 1, pp. 37-56.

Hartmann, H., Sabass, H., Tagivakatini, S. and Ries, F. (2010), *Climate Change Education in the Pacific Island States*, SPC-GIZ, Suva.

Heyd, T. and Brooks, N. (2009), "Exploring cultural dimensions of adaptation to climate change", in Adger, W.N., Lorenzoni, I. and O'Brien, K.L. (Eds), *Adapting to Climate Change: Thresholds, Values, Governance*, 1st ed., University Press, Cambridge, Cambridge, MA, pp. 269-282.

Hulme, M. (2016), *Weathered Cultures of Climate*, SAGE Publications, London.

Iati, I. (2008), "The potential of civil society in climate change adaptation strategies", *Political Science*, Vol. 60 No. 1, pp. 19-30.

IPCC (2013), *Climate Change 2013: The Physical Science Basis: Working Group I Contribution to the Fifth Assessment Report of the Intergovernmental Panel on Climate Change*, Stocker, T.F., Qin, D., Plattner, G.-K., Tignor, M., Allen, S.K., Boschung, J., Nauels, A., Xia, Y., Bex, V. and Midgley, P.M. (Eds), Cambridge University Press, Cambridge, MA.

Jones, R., Hennessy, K. and Page, C. (1999), *An Analysis of the Effects of Kyoto Protocol on Pacific Island Countries, Part 2: Regional Climate Change Scenarios and Risk Assessment Methods*, 1st ed., SPREP, Apia.

Kelman, I. and Khan, S. (2013), "Progressive climate change and disasters: Island perspectives", *Natural Hazards*, Vol. 69 No. 1, pp. 1131-1136.

Kelman, I., Gaillard, J.C. and Mercer, J. (2015), "Climate change's role in disaster risk reduction's future: beyond vulnerability and resilience", *International Journal of Disaster Risk Science*, Vol. 6 No. 1, pp. 21-27.

Kelman, I., Lewis, J., Gaillard, J.C. and Mercer, J. (2011), "Participatory action research for dealing with disasters on Islands", *Island Studies Journal*, Vol. 6 No. 1, pp. 59-86.

Korovulavula, I. (2016) "Climate change impacts in Ra and Kadavu provinces", *SPC Pacific Community Report*, available at: www.spc.int/wp-content/uploads/2016/12/Climate-change-impacts-Ra-Kadavu-Provinces-Fiji.pdf

Kunreuther, H. and Slovic, P. (1996), "Science, values, and risk", *The Annals of the American Academy of Political and Social Science*, Vol. 545 No. 1, pp. 116-125.

Lata, S. and Nunn, P. (2012), "Misperceptions of climate-change risk as barriers to climate-change adaptation: a case study from the Rewa Delta, Fiji", *Climatic Change*, Vol. 110 Nos 1/2, pp. 169-186.

Lefale, P.F. (2010), "Ua 'afa le Aso stormy weather today: traditional ecological knowledge of weather and climate: the Samoa experience", *Climatic Change*, Vol. 100 No. 2, pp. 317-335.

Leiserowitz, A. (2006), "Climate change risk perception and policy preferences: the role of affect, imagery, and values", *Climatic Change*, Vol. 77 Nos 1/2, pp. 45-72.

Lindell, M. and Perry, R. (2003), *Communicating Environmental Risk in Multiethnic Communities*, 1st ed., Sage, London.

McLeman, R. and Smit, B. (2006), "Migration as an adaptation to climate change", *Climatic Change*, Vol. 76 Nos 1/2, pp. 31-53.

Mcnamara, K.E. (2013), "Taking stock of community-based climate-change adaptation projects in the Pacific", *Asia Pacific Viewpoint*, Vol. 54 No. 3, pp. 398-405.

Mataki, M., Koshy, K. and Nair, V. (2006), "Implementing climate change adaptation in the Pacific Islands: adapting to present climate variability and extreme weather events in Navua (Fiji)", AIACC Working Assessments of Impacts and Adaptations to Climate Change, No. 34.

Méheux, K., Dominey-Howes, D. and Lloyd, K. (2007), "Natural hazard impacts in small Island developing states: a review of current knowledge and future research needs", *Natural Hazards*, Vol. 40 No. 2, pp. 429-446.

Mercer, J. (2010), "Disaster risk reduction or climate change adaptation: are we reinventing the wheel?", *Journal of International Development*, Vol. 22 No. 2, pp. 247-264.

Mercer, J., Kelman, I., Alfthan, B. and Kurvits, T. (2012), "Ecosystem-based adaptation to climate change in Caribbean small Island developing states: integrating local and external knowledge", *Sustainability*, Vol. 4 No. 8, pp. 1908-1932.

Ministry of Natural Resources and the Environment –MNRE (2013), "Current and future climate of Samoa: the Pacific Climate Change Science Program", available at: www.pacificclimatechangescience.org/wp-content/uploads/2013/06/3_PCCSP_Samoa_8pp.pdf (accessed 5 May 2017).

Mortreux, C. and Barnett, J. (2009), "Climate change, migration and adaptation in Funafuti, Tuvalu", *Global Environmental Change*, Vol. 19 No. 1, pp. 105-112.

Nakashima, D., Galloway, M., Thulstrup, H., Castilho, R. and Rubis, J. (2012), *Weathering Uncertainty: Traditional Knowledge for Climate Change Assessment and Adaptation*, UNESCO, Jakarta.

Nunn, P.D. (2009), *Understanding Environmental Decision Making in the Rural Pacific Islands*, Suva.

Nunn, P.D. (2010), "Bridging the gulf between science and society: imperatives for minimizing societal disruption from climate change in the Pacific", *Adaptation and Mitigation Strategies for Climate Change*, Springer Japan, Tokyo, pp. 233-248.

Nunn, P.D. (2012), "Climate change and Pacific Island countries, report background papers series 2012/07", *United Nations Development Programme*, Asia Pacific Regional Centre, Suva.

Nurse, L.A., Mclean, R.F., Agard, J., Briguglio, L.P., Duvat-Magnan, V., Pelesikoti, N., Tompkins, E. and Webb, A. (2014), "Small Islands, Part B: regional aspects", *Contribution of Working Group II to the Fifth Assessment Report of the Intergovernmental Panel on Climate Change*.

Orłowska, J. (2016), "Living on the sinking Islands", Social Aspects of Climate Change on Example of Maldives, PhD thesis, Institute of Philosophy and Sociology Polish Academy of Sciences, Warsaw.

Palmer, J.F. (1983), "Assessment of Coastal Wetlands in Dennis, Massachusetts", in Osmun, A. (Ed.), *The Future of Wetlands, Assessing Visual-Cultural Values*, Allanheld Osmun, Totowa, NJ.

Papoutsaki, E. and Harris, U.S. (2008), *South Pacific Islands Communication: Regional Perspectives*, University of the South Pacific Press, Suva.

Paton, K. and Fairbairn-Dunlop, P. (2010), "Listening to local voices: Tuvaluans respond to climate change", *Local Environment*, Vol. 15 No. 7, pp. 687-698.

Patt, A. and Schroter, D. (2008), "Perception of climate risk in Mozambique: implication for the success of adaptation", *Global Environmental Change*, Vol. 18 No. 1, pp. 458-467.

Pitt, D.G. and Sube, E.H. (1979), "The Q-sort method: use in landscape assessment research and landscape planning", *Proceedings of National Landscape: A Conference on Applied Techniques for Analysis and Management of the Visual Resource*, Vol. 1, pp. 227-234.

Robbins, P. and Krueger, R. (2000), "Beyond bias? The promise and limits of Q method in human geography", *Professional Geographer*, Vol. 52 No. 4, pp. 636-648.

Rubow, C. (2013), "Enacting cyclones: the mixed response to climate change in the Cook Islands", in Hastrup, K. and Skrydstrup, M. (Eds), *The Social Life of Climate Change Models*, 1st ed., Vol. 8, Routledge, Routledge Studies in Anthropology, Abingdon, pp. 57-76.

Rudiak-Gould, P. (2013), *Climate Change and Tradition in a Small Island State: The Rising Tide*, Routledge, Abingdon.

Ruosteenoja, K., Carter, T.R. and Tuomenvirta, H. (2003), "Future climate in world regions: intercomparison of model based projections for the new IPCC emissions scenarios", *The Finnish Environment*, Vol. 1 No. 644, p. 81.

Secretariat of the Pacific Community (SPC) (2012), *Pacific Platform for Disaster Risk Management*, SPC, Suva.

Secretariat of the Pacific Regional Environment Programme-SPREP (2009), "Pacific adaptation to climate change: Solomon Islands", *Report of In-Country Consultations*, pp. 1-37.

Tompkins, E.L. (2005), "Planning for climate change in small Islands: insights from national \hurricane preparedness in the Cayman Islands", *Global Environmental Change*, Vol. 15 No. 2, pp. 139-149.

UNESCO (2006), "Pacific education for sustainable development framework", Endorsed by the Pacific Ministers of Education, 27 September, Nadi.

UNESCO (2009), *Climate Frontlines*, UNESCO, Paris.

UNESCO (2014a), *Understanding Community Perceptions about Climate Change in the Pacific*, UNESCO, Apia.

UNESCO (2014b), "Institute for statistics (UIS) national education data", available at: www.epdc. org/country/samoa/search?school_level=82-53-59-113-50-48-118122&indicators=718&year_from= 2010&year_to=2017 (accessed 14 April 2017).

United Nations Development Program (2011), *Human Development Report 2011. Sustainability and Equity: A Better Future for All*, UNDP.

UNOHRLLS (2009), *The Impact of Climate Change on the Development Prospects of the Least Developed and Small Island Developing States*, 1st ed., UN-OHRLLS, Geneva.

Vanuatu Meteorology and Geo-hazard Department (VMGHD) (2011), "Current and future climate of Vanuatu: the Pacific climate change science program", available at: www. pacificclimatechangescience.org/wp-content/uploads/2013/06/15_PCCSP_Vanuatu_8pp.pdf (accessed 5 May 2017).

Vize, S. (2013), "Using education to bring climate change adaptation to Pacific communities", *Journal of Education for Sustainable Development*, Vol. 6 No. 2, pp. 219-235.

Wadley, G., Bumpus, A. and Green, R. (2014), "Citizen involvement in the design of technology for climate change adaptation projects in the Pacific", *Proceedings of the 26th Australian Computer-Human Interaction Conference on Designing Futures the Future of Design - OzCHI '14, ACM Press, New York, NY*, pp. 180-183.

Walid, M. (2017), "Towards a mega-Pacific Islands education curriculum for climate adaptation blending traditional knowledge in modern curriculum", in Filho, W.L. (Ed.), *Climate Change Adaptation in Pacific Countries*, 1st ed., Springer International, London, pp. 271-285.

Walsh, K.J.E., McInnes, K.L. and McBride, J.L. (2012), "Climate change impacts on tropical cyclones and extreme sea levels in the South Pacific – a regional assessment", *Global and Planetary Change, Elsevier B.V*, Vols 80/81, pp. 149-164.

Walshe, R.A. and Nunn, P.D. (2012), "Integration of indigenous knowledge and disaster risk reduction: a case study from Baie Martelli, Pentecost Island, Vanuatu", *International Journal of Disaster Risk Science*, Vol. 3 No. 4, pp. 185-194.

Wiese, V. (2012), "Education and climate change: living and learning in interesting times", *Journal of Peace Education*, Vol. 9 No. 2, pp. 321-323.

Further reading

Farbotko, C. and Lazrus, H. (2012), "The first climate refugees? Contesting global narratives of climate change in Tuvalu", *Global Environmental Change*, Vol. 22 No. 2, pp. 382-390.

Nunn, P.D., Runman, J., Falanruw, M. and Kumar, R. (2016), "Culturally grounded responses to coastal change on Islands in the federated states of Micronesia, Northwest Pacific ocean", *Regional Environmental Change*, Vol. 1 No. 13, pp. 428-439.

About the authors

Rory A. Walshe is a PhD Student at King's College London, Department of Geography and the University College London (UCL) Institute of Risk and Disaster Reduction (IRDR). His research interest is community resilience and vulnerability on small island developing states. This particularly includes the integration of local knowledge and culture within disaster risk reduction and climate change adaptation. His current research investigates contemporary and historical responses and perceptions of tropical cyclones in the Indian Ocean. Rory A. Walshe is the corresponding author and can be contacted at: rory.walshe@kcl.ac.uk

Denis Chang Seng currently works at the Intergovernmental Oceanographic Commission (IOC) of UNESCO. He is a Programme Specialist working jointly in the Tsunami Unit and the Ocean Observation and Services Section. In the tsunami unit, he is the Technical Secretary for the IOCs Intergovernmental Coordination Group for the Tsunami Early Warning and Mitigation System in the North-eastern Atlantic, the Mediterranean and connected seas (*ICG/NEAMTWS*). Previously, Denis was a UNESCO's Natural Science Programme Specialist for the Pacific Island States. He also worked at the United Nations – Institute for Environment and Human Security (UNU-EHS) in Bonn, Germany.

Adam Bumpus, BSc MA DPhil, is a Senior Lecturer in Environment, Innovation and Development, Faculty of Science, University of Melbourne, Australia. His work focuses on the business and policy pathways to a low carbon future including clean energy entrepreneurship, how mobile technologies can make local development more effective and how local sustainability innovations can become successful in the global economy.

Joelle Auffray is the Co-Founder and Director of Apidae Development Innovations Pty Ltd, a strategic communications and creative agency for sustainable development. Since 2012, Joelle has led climate change and sustainable development projects throughout the Asia-Pacific region, working with organisations such as UNESCO, UNDP, UNISDR, SPREP, SPC, ITTO, GGGI and the World Bank. Her expertise in community engagement, knowledge sharing, research and reporting and communicating program impact on sustainable development outcomes, has supported the work of intergovernmental organizations, government, donors and the private sector. She holds a master's degree in Global Media Communication at the University of Melbourne.

4

Mainstreaming climate change into the EIA procedures: a perspective from China

Qi Gao

KoGuan Law School, Shanghai Jiao Tong University, Shanghai, China

Abstract

Purpose – In the face of climate change, environmental impact assessment (EIA) and strategic environmental assessment (SEA) are expected to translate global or national mitigation and adaptation targets to project and plan levels of decision-making. This paper aims to examine how to transform China's EIA procedures to accommodate consideration of climate change and what constraints might be for doing so.

Design/methodology/approach – The main methodology used in this paper is doctrinal research, which is the primary legal methodology to find the law and interpret and analyse the document. Theoretical research is applied to analyse the ideas and assumptions of the mainstreaming approach. Comparative research is done to consider relevant international experiences.

Findings – Despite well-founded rationale for the mainstreaming approach, entrenched institutional, legal and technical obstacles cannot be neglected in the context of China. Urgent needs to fix existing EIA/SEA loopholes and improve the general enabling environment are also highlighted as a fundamental aspect of mainstreaming.

Originality/value – The potential of mainstreaming climate change into China's EIA procedures remains largely unexplored. As a ground-breaking work from China's perspective, the findings of this paper can serve as an important foundation for future research from legal and other perspectives.

Keywords Climate change, Mainstreaming, SEA, Climate feasibility study, EIA

1. Introduction

The probability of human influence in climate change has been raised by the IPCC (2013) from 90-95 to 95-100 per cent. Being the world's biggest CO_2 emitter (International Energy Agency, 2011), China released its "intended nationally determined contribution" (INDC) in June 2015 (National Development and Reform Commission, 2015). More recently, the Paris Agreement (Article 2) ambitiously declared its goal to keep global warming to "well below 2°C above pre-industrial levels". Nevertheless, specific means to implement and achieve such mitigation targets seems to be less of a concern for global discussion (Christopher, 2008). As the IPCC (2001) highlights interactions of climate change with other environmental stresses, one appealing suggestion is to translate global or national reduction targets to plan and project levels of decision-making through strategic environmental assessment (SEA) and environmental impact assessment (EIA), which have been widely used to address traditional environmental concerns.

Recent years have also witnessed an increasing emphasis on adaptation to build resilience to climate change related damages (Farber, 2011). Global attention has particularly focused on adaptation in developing countries, as they suffer the most and first from climate change. Being the largest developing country, China is very vulnerable to climate change (IPCC, 2014; National Development and Reform Commission, 2013). Article 7 of the Paris Agreement requires countries to strengthen institutional arrangements to support the synthesis of adaptation related information and knowledge. Adaptation strategies should be considered at various decision-making levels to reduce climate sensitive risks (Article 7). From this perspective, it is argued that the EIA procedures (hereinafter refers to both EIA and SEA) provide potential tools to collect and evaluate information on what to adapt to and how to adapt. This is in line with the "mainstreaming" approach to build resilience to climate change.

By nature, however, EIA is probably not suitable as a primary means to tackle climate change (Christopher, 2008). Although the EIA procedures could be used to mainly ensure *future* plans and projects not to significantly increase greenhouse gas (GHG) emissions or to better adapt to climate change, more comprehensive strategies are needed to deal with *existing* emission sources and climate sensitive sectors and projects (Christopher, 2008). But as a secondary tool, the potential of mainstreaming climate change into China's EIA procedures remains largely unexplored.

This paper aims to examine how to transform China's EIA procedures to accommodate consideration of climate change and what constraints might be for doing so. Both the potentials and boundaries of the EIA procedures are essential in understanding how they may incorporate climate change issues (Christopher, 2008). We begin with a theoretical analysis on the rationale for integrated consideration of climate change via the EIA procedures. Attention will then be paid to the existing climate feasibility study in China as a separate initiative to evaluate climate change impacts. In light of the progress and problems revealed in relevant legislation, key issues concerning the transformation of China's EIA procedures will be discussed in Section 4. Comparative research is also applied to consider relevant international experiences. The European Union (EU) and several countries (represented by Canada and Australia) have pioneered in this area. But the level of progress varies considerably among them (Agrawala *et al.*, 2011). This is followed by an examination of possible obstacles towards integrated consideration. Conclusions will be drawn accordingly in the context of China.

Unfortunately, due to the length limitation and a lack of transparency in relevant decision-making process[1], this paper is unable to provide a detailed first-hand quantitative analysis. Nevertheless, as a groundbreaking work, barriers and potentials identified in this paper can serve as an important foundation for future research from legal and other perspectives.

2. Rationale for integrated consideration of climate change via the environmental impact assessment procedures

As a means that has been widely applied since the late 1990s to deal with gender inequality, AIDS and environmental degradation in developing countries (Oates *et al.*, 2011), the mainstreaming approach is of increasing significance in the context of climate change. It generally indicates that "cross-cutting issues should influence the 'mainstream' activities of development, rather than being addressed in separate initiatives" (Oates *et al.*, 2011). This is in contrast with the fact that many climate change related decisions "continue to be taken with little or no regard to climate change" (Kok and Coninck, 2007). Proponents states that the aim of mainstreaming is "to capture the potential in other policy areas and sectors for implementing climate-friendly and climate-safe development pathways" (Kok and Coninck, 2007).

In past research and practice, the mainstreaming approach has received more recognition with respect to climate change adaptation than mitigation, although it is in fact necessary for both (Kok and Coninck, 2007). With regard to adaptation, poverty alleviation, rural development, agriculture and disaster management are identified as major areas that require mainstreaming (Kok and Coninck, 2007). For mitigation, issues like energy restructure and security, air quality, human health, transport, trade and finance should integrate climate change concerns (Kok and Coninck, 2007). Development in these areas involves a large number of projects and plans and the EIA procedures play a vital role in their decision-making processes. By incorporating climate change concerns into the EIA procedures, it is expected to:

> [...] increase the scientific understanding of climate-sensitive systems under changing climate conditions, to inform the specification of targets for the mitigation of climate change, to prioritize political and research efforts to particularly vulnerable sectors and regions, and to develop adaptation strategies that reduce climate sensitive risks independent of their attribution (Füssel and Klein, 2006).

As a fast-growing developing country, China has been challenged by both industrial and post-industrial environmental risks. The needs for climate change mitigation and adaptation, cleaner water and air are all justifiable and no one claim necessarily prevails over others[2]. Synergies are therefore identified and trade-offs are made among different environmental policy goals. These competing but overlapping needs should be evaluated and considered in an integrated manner. The EIA and SEA are supposed to fulfil such expectation at different decision-making levels. This is particularly referred to by the EU Guidance on Integrating Climate Change and Biodiversity into EIA (European Commission, 2013). The absent of consistent consideration could lead to poor environmental judgement in the age of climate change.

The biggest merit of the mainstreaming approach is probably that:

> [...] it can be implemented immediately, without a formal connection to the UNFCCC process, and in this way help the climate regime forward at a time that international negotiations on both climate adaptation and mitigation are at a crucial stage (Kok and Coninck, 2007).

Although the Paris Agreement finally entered into force in November 2016, a shadow has been casted on political will for international cooperation by President Trump's decision to withdraw from the agreement (Shear, 2017). Moreover, the Paris Agreement itself mainly serves as a framework for further cooperation rather than a strict action plan. It does not entail any legally binding emission reduction obligations. Instead, it is the INDCs determined by each participating nation that constitute the basis for mitigation (Article 4). Under such circumstances, the mainstreaming approach is a practical and meaningful way for countries to act on climate change in light of different national circumstances.

3. Climate feasibility study in China: a separate initiative
In contrast to the mainstreaming approach, China has adopted an alternative "stand-alone" tool to evaluate and adapt to climate change impacts. Early in the end of 1999, China's Meteorology Law required certain kinds of plans and projects to go through the so-called "climate feasibility study". The Management Measures on Climate Feasibility Study was then promulgated by the China Meteorological Administration in 2008 to provide detailed requirements.

Only certain kinds of plans and projects "closely related to climate conditions" are subject to climate feasibility study (Article 4)[3]. As can be discerned in Section 4, the range of plans and projects subject to climate feasibility study is more restricted than that of the EIA procedures in China. Moreover, the current regulation fails to provide further explanation on the terms of "important", "major" and "closely related to climate conditions". The discretion is completely left to decision makers, thus giving leeway to bypass the process. In contrast, a mandatory list is adopted in China to specify the kinds of projects that are subject to EIA, and all sectorial and comprehensive plans proposed by governments above the municipal levels are subject to SEA.

The Management Measures (Article 1) alleges to take both climate change mitigation and adaptation into consideration. In fact, based on the title and content of the study, it is the feasibility of proposed activities under certain climate conditions that is the focus of the study. With regard to mitigation, it only considers the impact of plans and projects on local climate[4]. This restriction denies the possibility of translating international or national targets on climate change mitigation into the local decision-making processes.

The Management Measures (Article 8) provides some general requirements on the scope of the study. For example, it should evaluate the probability of extreme weather events and the proposed plan/project's climate suitability, climate risk and potential impact on local climate. Suggestions to prevent or mitigate impacts should be provided accordingly. But the existing regulation fails to provide any operable guidelines regarding the content and extent of the matters to be covered by the study. As will be discussed in Section 4.4.2, Canada and the EU has developed more specific guidelines to facilitate the scoping process.

The design of the study is unable to cope with scientific uncertainty and complexity of climate change. Articles 8 and 9 of the Management Measures designate existing basic data or historical records as the main sources of information for the feasibility study. It reflects an inertia understanding of climate change based on ecological stationarity and scientific certainty. As the study is referred to as "climate" feasibility study rather than "climate change" feasibility study, it seems that it does not intend to or is unable to differentiate climate change with climate variability (Wang and Gao, 2015). Climate variability indicates a year-to-year fluctuation in the climate record and is generally predicable through historical records (He, 2013). According to Article 1(2) of the UNFCCC, however, climate change is "a change of climate [. . .] which is in addition to natural climate variability observed over comparable time periods". Drawing a clear line between climate change and climate variability could be difficult and controversial in practice, but not separating the two concepts is misleading.

The feasibility study should be conducted by qualified institutions. According to Article 5 of the Management Measures, it is the meteorological bureaus that are held responsible for conducting the study for proposed plans. With regard to proposed projects, the study should be done by institutions that have obtained relevant qualification from the China Meteorological Administration (Article 7). A specific Interim Management Measures has been issued by the China Meteorological Administration in 2009. But it is not even as sophisticated as the requirements on EIA qualifications, especially with regard to liability and accountability[5]. In practice, qualifications are usually granted to climate centres established under the China Meteorological Administration and meteorological bureaus at the provincial level. The monopoly status and quasi-administrative feature of climate centres raise genuine concerns for high risk of rent seeking, local protectionism and manipulation of the study.

Article 5 of the Management Measures requires plan-making authorities to take full account of the conclusions of the feasibility study. But it fails to provide any further procedural guidance on what constitutes as adequate consideration. The Management Measures is merely a department rule, the hierarchy of which is the lowest in China, and it often has little impact on other government sectors. In addition, the meteorological bureaus in China are not powerful compared to other departments relevant to climate change. The situation is slightly different with regard to proposed activities, the feasibility study report of which should be reviewed and authorized by the meteorological bureaus at either national or provincial levels (Article 11). Given the close linkage between climate centres qualified to conduct the study and meteorological bureaus responsible for issuing such permits, the reliability of this procedure is also questionable, to say the least.

Public participation is considered as a valuable mechanism to improve the quality of environmental decision making. Unfortunately, the Management Measures fails to make any arrangements for information transparency and public participation. In contrast, EIA represents a comparatively good example in China for protecting the public's right to participate, although problems still abound in detailed arrangements and implementation (Gao, 2014).

In addition, ecological non-stationarity and scientific uncertainty of climate change justify the necessity to highlight monitoring and follow-up analysis. It is mandatory for the EIA procedures in China, but it is not even briefly mentioned in the Management Measures.

Finally, administrative penalties for violation of the Management Measures cannot generate incentives for compliance. In case of severe non-compliance, such as failing to conduct the feasibility study or committing data forgery, a maximum fine of merely 30,000 yuan is applicable. Combined with the previously mentioned deficits, it can be concluded that climate feasibility study is more window dressing than a substantive tool for climate change evaluation.

As a separate initiative, climate feasibility study completely alienates climate change from traditional environmental concerns. According to Shackley and Wynne (1995):

> [S]ingling out climate change as the prism for investigating systems is a classic form of reductionism since it assumes, *a priori*, that the processes, impacts and responses of relevance in a particular system are indeed those collected under the banner of climate change. Other possible variables are simplified and backgrounded: they are not the key variables of interest however much their interactive importance at a secondary level may be recognized.

Therefore, the following discussion will explore the possibility of incorporating climate change into China's EIA procedures.

4. Key issues on mainstreaming climate change into China's environmental impact assessment procedures

Early in 1997, scholars have envisioned a "conceptual framework for considering integrated assessments", which is characterized by nonlinear, adaptive and participatory analysis, pluralistic consideration of alternatives and a combination of quantitative and qualitative methodologies (Rothman and Robinson, 1997; Füssel and Klein, 2006). Progress in reality, however, has been really slow.

Although several governments have indicated their willingness to move in this direction, only a few of them (represented by Canada) actually developed operational guidelines and/ or adjusted regulatory frameworks (Agrawala *et al.*, 2011). Breakthrough progress occurred between 2013 and 2014, during which the Directive 2014/52/EU was adopted to formally incorporate climate change mitigation and adaptation concerns into EIA (Annexes III(1)(f), IV(5)(f)) and a specific EU Guidelines was released to provide further instructions for its member states (European Commission, 2013). Therefore, all EU member states are now subject to a compulsory obligation to consistently consider climate change with other environmental concerns at the project level.

EIA has been one of the key environmental regulatory tools in China for decades. Currently, the EIA Law (2016) and the SEA Regulation (2008) serve as the key legal documents in this area. Moreover, the Ministry of Environmental Protection (MEP) in China has released several technical guidelines for EIA and SEA. In light of comparative experiences and the status quo of the EIA procedures in China, the following discussion will focus on several key issues affecting the operationalization of an integrated consideration in China.

4.1 The precautionary principle and scientific uncertainty

EIA has been recognized as an important preventive mechanism since its origination. It "rests on cumulative experience concerning the degree of risk posed by an activity" and presupposes "the notion of an objective assessment of risks in order to reduce the probability of their occurrence" (Sadeleer, 2002). EIA is therefore not initially designed to deal with "surprises". Unfortunately, we are currently living in a risk society where surprises abound. Under such circumstances, the EIA itself should be redesigned to deal with such uncertainties in accordance to the precautionary principle, which calls for actions to avoid or diminish unacceptable harm that is scientifically plausible but uncertain (Sadeleer, 2002). Correspondingly, the Directive 2014/52/EU (the EIA Directive up to date) explicitly recognizes the precautionary principle in its preambles.

According to Craig (2010), the assumptions of ecological stationarity fail to "fit a world of continual, unpredictable, and nonlinear transformations of complex ecosystems". It is further pointed out that: "[c]limate change science is unlikely ever to be able to present climate information in the same manner as historical data to which the EA community is accustomed" (Agrawala et al., 2011). Hence, a qualitative approach is preferred rather than relying on quantitative data to predict impact (European Commission, 2013; Sadeleer, 2002). Some basic approaches are further recommended for addressing climate change uncertainties in EIA, such as scenario analysis, sensitivity analysis and probabilistic analysis (Byer and Yeomans, 2007; European Commission, 2013). Public perceptions and lay knowledge are also highlighted to provide a more comprehensive understanding of complex scientific situations (Sadeleer, 2002).

The application of the precautionary principle also requires a more dynamic and learning-oriented decision-making process. Repeated EIA at regular intervals can help public authorities adapt their decisions to new results (Sadeleer, 2002). This is reflected in the adoption of a post-project analysis in the 1991 Espoo Convention (Sadeleer, 2002). Hence, reversible options are also preferred to take advantage of the new knowledge (Sadeleer, 2002).

As revealed in Section 3, challenges posed by scientific uncertainty are not fully recognized by climate change regime in China. The situation is not that optimistic even in Canada. Empirical research demonstrates that many cases of EIA that claim to consider climate change in fact rely on historical climate data at least for some aspects of evaluation (Agrawala et al., 2011). This is attributable to "the difficulties for EIA practitioners to address climate change information and its uncertainties" and "the low accessibility for EIA practitioners to user-friendly climate change information" (Agrawala et al., 2011).

If China did attempt to introduce climate change factors into the EIA procedures, relevant information gaps and technical difficulties will unlikely to be any easier for EIA practitioners in China, which will be discussed in detail later. Moreover, the existing national legislation in China has not explicitly recognized the precautionary principle, although

Wang (2014) broadly interprets the principle of prevention (Article 5 of the 2014 Environmental Protection Law) to include the meaning of precaution.

4.2 Mitigation aspects and adaptation aspects

Traditionally, the EIA procedures are designed to evaluate and prevent a significant impact of proposed actions on environment. With regard to climate change, it means that a proposed action's impact on GHG emissions (e.g. increased emissions or sinks of GHG) should be examined. In addition to this natural extension of EIA, mainstreaming proponents further argue that the EIA procedures should consider the impact of climate change on proposed actions.

It should be noted that possible synergies and conflicts exist between climate change mitigation and adaptation needs (European Commission, 2013). Füssel and Klein (2006) also highlight "differences in the typical temporal and spatial scales at which mitigation and adaptation take place and in their respective information needs". Due to the nature of a proposed activity, the actual evaluation process may focus on one aspect of climate change than the other (Federal-Provincial-Territorial Committee on Climate Change and Environmental Assessment, 2003).

Empirical study shows divergent practices. Canada and the EU are known for taking a holistic approach to incorporate both aspects of climate change into EIA (Federal-Provincial-Territorial Committee on Climate Change and Environmental Assessment, 2003; European Commission, 2013). Small Island Developing States, on the other hand, tend to use EIA as a tool to adapt to climate change (Agrawala et al., 2011). This divergence can be attributable to priority differences, for the latter are known to be very vulnerable to climate change.

As analyzed in Section 3, the current regulation and practice of China's climate feasibility study tends to emphasis on the vulnerability of a proposed activity to climate change. Being the world's biggest CO_2 emitter (International Energy Agency, 2011) and a rapidly growing economy, a lack of operable mechanisms to evaluate the impact of a proposed action on climate change will hinder the implementation of China's INDC and its progress evaluation under the Paris Agreement.

4.3 Integrated consideration of climate change at the levels of environmental impact assessment and/or strategic environmental assessment

Climate change mitigation and adaptation can be better considered at what levels of EIA is another question put in front of mainstreaming proponents. Compared to well-developed EIA process for specific activities, SEA is still a relatively new concept, especially to developing countries. At the time of writing, China only applies SEA to certain kinds of plans, while policies are not subject to SEA.

Some argue that efforts would be better spent on integrating climate change into SEA (Christopher, 2008). It is noted that plans usually have a more profound and cumulative impact on climate change mitigation and adaptation than site-specific projects (Christopher, 2008). Plan making is also crucial in translating national mitigation and adaptation targets in respective sectors, such as transportation, energy and agriculture (Wende et al., 2012). But others point out that climate change itself "is a result of a series of individually small actions, and the 'solution'- if there is one- will likely result from a series of separate steps" (Christopher, 2008). Even if GHG emissions are more difficult to evaluate at the project level, specific measures to adapt to climate change are certainly required for individual projects.

Among the few pioneers that actually implement on mainstreaming climate change into the EIA procedures, The Netherlands integrates climate change into SEA, while Canada and Australia incorporates climate change into EIA (Modak and Ginoya, 2013). The EU

Guidelines recognizes the necessity to incorporate climate change at both project and strategic levels of assessment (European Commission, 2013), but so far the SEA Directive has not been amended to reflect changes toward this direction.

While there are merits on both sides theoretically, challenges in reality will likely to make a difference in the process of developing operational guidelines, such as access to climate change information, availability of reliable climate change analysis models and the performance record of EIA and SEA. As will be discussed later, EIA is relatively more advanced than SEA in China, and it imposes more obligations on developers rather than governments. If China did decide to mainstream climate change into the EIA procedures, there is less resistance at the level of EIA than SEA. This is in line with the process of relevant legal reform at the EU level.

4.4 Potential entry points for considering climate change via the environmental impact assessment procedures

Mainstreaming climate change into the EIA procedures requires proper adjustments on the existing mechanisms to bridge the gap between intention and action. This is reflected in the introduction of climate change parameters, methodologies and thresholds; increasing significance of certain EIA/SEA phases; and urgent needs to fix existing EIA/SEA loopholes. Given the complexity of many issues involved, only a few countries and the EU have developed specific guidelines to facilitate relevant practice (Agrawala *et al.*, 2011; European Commission, 2013; Wentz, 2015).

4.4.1 Screening. The screening process determines whether an EIA or SEA process will be required for certain proposed actions (European Commission, 2009a). An EIA or SEA is usually necessary only when a proposed activity is likely to cause a "significant" adverse impact. The EU Directive 2011/92/EU (the EIA Directive) adopts two approaches to screen projects. All projects listed in Annex I are subject to EIA. Projects listed in Annex II are left to the discretion of EU member states, which can be decided either through a case-by-case examination or according to criteria set by the member states. Annex III then set out the selection criteria that should be applied to the case-by-case examination and threshold setting.

Proponents suggest that the criteria for threshold setting should be readjusted to screen proposed activities that are likely "to have significant effects on, or be significantly affected by climate change" (European Commission, 2013)? Drawing a line here, however, is a controversial judgment. Some argue that the threshold should be set based on "the proportion of global anthropogenic GHG emission that can be attributed to a particular action" (Wentz, 2015). But the nature of climate change challenge makes it impossible to attribute climate impacts to any single action (Wentz, 2015).

In a revised draft guidance of Council on Environmental Quality in the USA, a reference point of 25,000 metric tons of CO_2 equivalent annually is recommended as a threshold for quantitative analysis of GHG emissions (Wentz, 2015). But it does not clarify on whether this benchmark applies to both direct and indirect emissions (Wentz, 2015). It also fails to explain why other parameters (e.g. carbon sinks and emission intensity) are not accepted as an independent factor for screening. The modules used to quantify GHG emissions can be subject to disputes as well. With regard to climate change impact on a proposed activity, the draft guidance provides no specific benchmark, but it lists a range of impacts that should be incorporated in assessment (Wentz, 2015). Namely, climate change adaptation is considered as an issue for scoping.

The EU Guidelines highlights the importance of incorporating climate change into environmental assessment "at an early stage (screening and scoping)", but the screening

environmental protection authorities and traditional EIA practitioners to evaluate climate change impacts. Under such circumstances, cooperation from meteorological bureaus and climate centers remains crucial in practice, at least in a short term. As the EIA service market matures and institutional reform deepens in China, there may be a possibility that the situation could be gradually improved in the future. Attention should be paid to imbalanced development of meteorological infrastructure and expertise across China as well.

Moreover, future guidelines on integrated evaluation should try to separate climate change and climate variability to consider scientific uncertainty that underlies climate change concerns. Information and technical gaps will therefore be revealed and then improvement efforts can be made accordingly.

As mentioned in the beginning of Section 4, integrated assessment in the age of climate change is featured with pluralistic consideration of alternatives. Christopher (2008) also highlights the potential of utilizing alternative analysis in the USA to incorporate climate change considerations. This opportunity, however, barely exist in China where alternatives are not a compulsory part of the EIA procedures.

4.4.4 Public participation. As the international debate on climate change has been distant from local public, the EIA procedures are expected to provide valuable opportunities for the public to participate in climate change debate and knowledge expansion on a case-by-case basis. In light of scientific uncertainty and compounded environmental stresses, public participation is crucially needed to increase the accountability of decision-making to interest groups (Sadeleer, 2002) and hence provide legitimacy to controversial development decisions. Effective public participation could also contribute to better understanding of complex scientific situations and developing suitable measures to mitigate or adapt (Agrawala *et al.*, 2011; Sadeleer, 2002).

Necessary transparency and public awareness are considered as key aspects of capacity building in Article 11 of the Paris Agreement. Article 12 further requires countries to take measures to enhance public access to information and public participation. China's EIA procedures have a relatively better record in this aspect than climate feasibility study. But there remains a long way to go before it could meet the above expectation in practice.

In the USA, public participation is considered as an important approach to introduce climate change concerns into the EIA process even when the USA is not subject to any mandatory emission reduction targets (Christopher, 2008). In China, however, climate change is not a priority concern for the general public compared to other environmental problems, such as PM 2.5 pollution. Although civil society development is gaining momentum in China, many legal and institutional restrictions on the establishment and operation of NGOs still exist (Gao, 2014), which considerably lower the capacity of NGOs to actively participate.

Moreover, existing legislation on public participation in China's EIA procedures is still problematic. In a nutshell, public participation in the SEA process suffers from limited opportunities for the public to participate in SEA[6], restricted access to information, late introduction of public participation, no detailed arrangements on participation procedures and the absence of judicial remedy (Gao, 2014).

The EIA process is the only mechanism in China that has developed a relatively comprehensive legal framework to facilitate public participation (Gao, 2014). Even so, public participation in China's EIA process remains to be highly controlled by developers and governments. Gao (2014) repeatedly criticizes the lack of requirements for participation at earlier stages of EIA[7], insufficient information disclosure, short minimum time-frame set for public participation (10 days), inappropriate approaches applied to seek public opinions[8] and insufficient requirements on environmental protection authorities to consult public during the review of EIA reports[9]. As a result, the public is often unwilling to actively participate in a window dressing procedure.

Under such circumstances, to what extent public participation can contribute to mainstreaming climate change into China's EIA procedures in practice is highly doubtful. Climate change response not only requires renovation of traditional environmental law but also demands better design and implementation of many existing environmental mechanisms and the general improvement of enabling environment. In this sense, arguments in favour of mainstreaming climate change into the EIA procedures will add momentum to further enhancement of public participation in China.

4.4.5 Monitoring and follow-up analysis. In the context of scientific certainty, traditional EIA procedures build on the idea of taking *ex ante* measures to prevent the occurrence of environmental problems. Hence, a follow-up analysis is often not mandatory in many EIA procedures across the world. In contrast, a more dynamic, learning oriented decision adjustment process is recommended to evaluate climate change impacts (Ruhl, 2010; Farber, 2009). Through monitoring and follow-up analysis, the initial projections of climate change can be verified or modified. The flexibility to change decisions is crucial in case the predictions were incorrect or new information became available (European Commission, 2013; Sadeleer, 2002).

According to Articles 15 and 27 of the EIA Law, a follow-up analysis during and/or after the implementation of proposed plans and projects is mandatory, and measures shall be taken to mitigate environmental impacts[10]. Moreover, environmental protection authorities should carry out a follow up inspection on environmental impacts of a proposed project (Article 28). But no liability is stipulated for violation; hence, the compliance of follow-up analysis is not guaranteed. As will be revealed in Section 5, in a country where the EIA procedures have a poor reputation in general, one should not assume that the follow up analysis can function properly.

5. Obstacles for mainstreaming climate change into China's environmental impact assessment procedures

5.1 Institutional barriers

The mainstreaming approach indicates that "actors whose main tasks are not directly concerned with mitigation of, or adaptation to, climate change also work to attain these goals" (European Commission, 2015). The introduction of climate change discourse is therefore expected to promote cooperation among government sectors. In China, however, climate change regulation has to struggle with, if not aggravates, deep rooted institutional barriers and bureaucratic infighting among different government sectors (Wang and Gao, 2015).

According to Article 10 of the Environmental Protection Law, the MEP is in charge of China's environmental protection, but in reality, it often fails fulfil such expectation. With regard to climate change, it is the National Department and Reform Commission (NDRC) that plays a dominate role in decision making (Wang and Gao, 2015). To some extent, this is a practical and preferable choice. Climate change response calls for profound economic reform and the NDRC is a macroeconomic management agency that enjoys broad administrative and planning control over the Chinese economy. As a much more powerful institution, the NDRC is arguably better equipped to translate national mitigation and adaptation targets to different levels of planning.

The downside of this administrative arrangement is that climate change regulation is largely isolated from environmental governance regime (Wang and Gao, 2015). A holistic consideration of climate change and other environmental concerns will lead to power expansion of the MEP, which is very controversial due to entrenched bureaucratic interests.

In nature, however, climate change is a part of the bigger picture of environmental protection. It reshapes the scientific basis of environmental protection and natural preservation (Wang and Gao, 2015). In addition, effective climate change response still relies on the improvement of existing environmental legislation and its implementation (Wang and Gao, 2015). But the prospect of institutional reform towards this direction is not optimistic at the moment.

5.2 Challenges on technology and resources
Climate change entails a degree of scientific uncertainty that is unaccustomed to the existing environmental regulatory mechanisms. Disputes concerning methodologies, parameters, models and technologies applied for integrated assessment have been highlighted by past empirical research (Agrawala *et al.*, 2011). This is one of the main reasons for the majority of EU countries not to evaluate climate change impacts via the EIA procedures before the revision of the EIA Directive in 2014 (European Commission, 2009b). Nowadays, challenges are still highlighted due to the long-term and cumulative nature of impacts, complexity of cause–effect relationship and scientific uncertainty (European Commission, 2013).

China currently fails short of a strong scientific background to support consistent climate change assessment on a frequent basis. The availability and reliability of necessary information also affect the accuracy of integrated assessment. The above obstacles are particularly challenging for less developed regions. Unfortunately, these areas have been targeted by many controversial development projects and plans.

A lack of necessary financial and human resources and unbalanced distribution of limited resources further aggravate the prospect for integrated consideration. Climate change assessment via the EIA procedures entails additional work compared to traditional EIA or SEA (Agrawala *et al.*, 2011). Due to the immaturity of EIA service market, existing institutional barriers and unbalanced regional development, one can safely predict that the provision of necessary resources is likely to be insufficient in China.

5.3 Malfunction of the environmental impact assessment procedures in China
The track record of the EIA procedures in China is far from reassuring, even absent the complications generated by climate change. It is therefore concerned that mainstreaming climate change into the EIA procedures could be too much to ask for an overstretched, malfunctioned system.

The ineffectiveness of the SEA procedure in China can be attributed to the problems of its own design and deficiencies of the environmental administrative regime. Main problems include, but are not limited to, a lack of alternative consideration, insufficient requirements on information disclosure and public participation, deficient noncompliance and liability requirements[11], restricted government budget for conducting SEA, and a relatively weak status of environmental protection authorities against other governmental sectors (Gao, 2014).

EIA is relatively more sophisticated than SEA in China. This is partly due to the fact that obligations to conduct EIA in China are mainly applied to developers, instead of the governments responsible for decision-making. Hence, there are less political barriers for adopting more detailed and stringent requirements. But problems suffered by SEA also exist to varying degrees in EIA (Gao, 2014). Recent years have witnessed some reform measures to improve the effectiveness of EIA, represented by the revision of the Environmental Protection Law in 2014 to require the disclosure of full EIA report (Article 56), the reform to separate the EIA agencies that used to be subordinate to environmental protection authorities (Liu, 2015) and the revision of the EIA Law in 2016 to increase punishment for failure to conduct EIA (Article 31). But the effectiveness of these measures remains to be seen.

Given the unsatisfactory record of the EIA procedures in China, a difficult question is as follows: if the existing regulatory regime has failed to perform effectively, is the mainstreaming approach necessarily more preferable than separate initiatives when dealing with cross-cutting issues?

6. Conclusions

With increasing pressure from climate change and growing momentum on mainstreaming, China now faces a stark dilemma regarding how to bridge between distant global aspirations and domestic decisions. The mainstreaming proponents argue that climate change can be addressed at smaller scales via the EIA procedures. A holistic consideration of climate change and other environmental protection goals is ideal for dealing with potential synergies and trade-offs.

But notable challenges and difficulties for doing so pose serious questions on the feasibility and effectiveness of its implementation in China. China has used to take separate initiatives on dealing with climate change. Political inertia and entrenched vested interests will likely to stagnate reforms towards integrated assessment. Combined with unresolved challenges on technology and resources and the existing deficiencies of the EIA procedures in China, the mainstreaming proponents seem to be unable to provide a realistic prospect to generate incentives for reforms.

At this stage, it is more likely that China will continue to address climate change issues in a more separate way. Nevertheless, it is valuable to at least put the other option on the table for discussion. The early stage of integrated consideration should not be the reason for rejection. Rather, it should serve a reasonable justification for more research and pilot practice. A learning by doing process should be applied to further explore and test the prospect of this innovative approach in China. Compared to climate change mitigation, adaptation at a plan or project level may be more attractive for policy makers at the moment.

It is hoped that the introduction of climate change discourse into the EIA procedures can generate more momentum on improving pre-existing deficiencies of the EIA mechanisms and the general enabling environment in China. This level of mainstreaming is so basic and fundamental that it tends to be overlooked in the process of climate change related reforms. As reforms at this scale usually involve many vested interest conflicts, it is worried that mainstreaming at this level may overload climate change discourse. But without progress in this sense, the achievement of mitigation and adaptation targets cannot be guaranteed in the long run.

Notes

1. For example, a full EIA report in China was not accessible for the general public until the revision of the Environmental Protection Law in 2014. The report of climate feasibility study remains inaccessible to the public in China until the day of writing.

2. Of course, priorities vary given different circumstances and preferences.

3. In particular, it refers to urban and rural plans, development and construction plans on crucial sectors or regions; important infrastructure projects, major public projects and large-scale construction projects; important regional economic development projects and major projects to restructure regional agriculture; large-scale development projects on wind and solar energy; other plans and projects required by law.

4. Local climates describe those climates which have influence over very small geographical areas, perhaps only a few miles or tens of miles across.

5. The qualification approval of EIA agencies has been subject to fierce criticism in China, and a reform has been initiated to tighten the management of EIA agencies (Ministry of Environmental Protection, 2013).

6. Public participation in China only applies to SEA of specific/sector-based plans. It is not a mandatory process during the SEA of comprehensive plans.

7. Public participation is generally introduced in the EIA process when the EIA report is almost finished.

8. The Interim Measures of Public Participation in EIA lists several approaches for public participation, ranging from questionaries to hearings. In reality, the public is often involved only via questionaries, which are mainly designed based on similar templates and can be easily manipulated (Chen, 2015).

9. Most requirements on public consultation apply to the developers rather than the environmental protection authorities.

10. In particular, a follow-up analysis is required for plans after implementation and for projects during their construction and operation processes.

11. The most common remedy is administrative punishment under the Civil Servants Law, and there is no access to justice for the public on the matter (Gao, 2014).

References

Agrawala, S., Kramer, A.M., Prudent-Richard, G. and Sainsbury, M. (2011), "Incorporating climate change impacts and adaptation in environmental impact assessments: opportunities and challenges", Environmental Working Paper No.24, OECD.

Byer, P.H. and Yeomans, J.S. (2007), "Methods for addressing climate change uncertainties in project environmental impact assessments", *Impact Assessment and Project Appraisal*, Vol. 25 No. 2, pp. 85-99.

Chen, J. (2015), "Forgery on public satisfaction regarding a project in Ningde, Fujian Province: the EIA agency claims no responsibility", *The Beijing News*, available at: www.bjnews.com.cn/news/2015/06/25/368277.html (in Chinese) (accessed 3 August 2017).

Christopher, C.W. (2008), "Success by a thousand cuts: the use of environmental impact assessment in addressing climate change", *Vermont Journal of Environmental Law*, Vol. 9 No. 3, pp. 549-613.

Craig, R.K. (2010), "Stationarity is dead'-long live transformation: five principles for climate change adaptation law", *Harvard Environmental Law Review*, Vol. 34, pp. 9-73.

European Commission (2009a), "Report from the Commission on the Application and Effectiveness of the EIA Directive (Directive 85/337/EEC, as Amended by Directives 97/11/EC and 2003/35/EC)", available at: http://eur-lex.europa.eu/legal-content/EN/TXT/PDF/?uri=CELEX:52009DC0378& from=EN (accessed 3 August 2017).

European Commission (2009b), *Study Concerning the Report on the Application and Effectiveness of the EIA Directive*, available at: http://ec.europa.eu/environment/archives/eia/pdf/eia_study_june_09.pdf (accessed 3 August 2017).

European Commission (2013), *Guidance on Integrating Climate Change and Biodiversity into Environmental Impact Assessment*, available at: http://ec.europa.eu/environment/eia/pdf/EIA%20Guidance.pdf (accessed 3 August 2017).

European Commission (2015), *Climate Policy Mainstreaming*, available at: http://ec.europa.eu/clima/policies/brief/mainstreaming/index_en.htm (accessed 3 December 2015).

Farber, D.A. (2009), "Adaptation planning and climate impact assessments: learning from NEPA's flaws", *The Environmental Law Reporter*, Vol. 39, pp. 10605-10614.

Farber, D.A. (2011), "The challenge of climate change adaptation: learning from national planning efforts in Britain, China and the USA", *Journal of Environmental Law*, Vol. 23 No. 3, pp. 359-382.

Federal-Provincial-Territorial Committee on Climate Change and Environmental Assessment (2003), *Incorporating Climate Change Considerations in Environmental Assessment: General Guidance for Practitioners*, Canadian Environmental Assessment Agency, available at: www.canada.ca/content/dam/canada/environmental-assessment-agency/migration/content/a/4/1/a41f45c5-1a79-44fa-9091-d251eee18322/incorporating_climate_change_considerations_in_environmental_assessment.pdf (accessed 3 August 2017).

Füssel, H.M. and Klein, R.J.T. (2006), "Climate change vulnerability assessments: an evolution of conceptual thinking", *Climate Change*, Vol. 75 No. 3, pp. 301-329.

Gao, Q. (2014), *A Procedural Framework for Transboundary Water Management in the Mekong River Basin: Shared Mekong for a Common Future*, Brill.

He, X. (2013), "Integrating climate change factors within China's environmental impact assessment legislation: new challenges and developments", *Law, Environment and Development Journal*, Vol. 9 No. 1, pp. 50-67.

International Energy Agency (2011), *CO2 Emissions from Fuel Combustion: Highlights*, 2011 ed., International Energy Agency, Paris.

IPCC (2001), *Climate Change 2001: Impacts, Adaptation, and Vulnerability*, WGII TAR, Cambridge University Press, Cambridge.

IPCC (2013), *Climate Change 2013: The Physical Science Basis (Summary for Policymakers)*, WGI AR5 SPM, IPCC.

IPCC (2014), *Climate Change 2014: Impacts, Adaptation, and Vulnerability*, WGII AR5, Cambridge University Press, Cambridge.

Kok, M.T.J. and Coninck, H.C. (2007), "Widening the scope of policies to address climate change: directions for mainstreaming", *Environmental Science and Policy*, Vol. 10 Nos 7/8, pp. 587-599.

Liu, Q. (2015), *China's Environmental Ministry Launches Anti-Graft Reforms*, Chinadialogue, available at: www.chinadialogue.net/article/show/single/en/7771-China-s-environment-ministry-launches-anti-graft-reforms (accessed 3 August 2017).

Ministry of Environmental Protection (2013), *To Promote the Reform of EIA Approval and the Functional Transformation, MEP Continuously Releases Three Documents to Delegate Approval Authorities, Intensify Information Disclosure, and Strengthen Supervision of EIA*, available at: http://english.mep.gov.cn/News_service/news_release/201312/t20131218_265214.htm (accessed 3 August 2017).

Ministry of Environmental Protection (2014), *Openly Seeking Opinions on the Catalogue of Construction Projects Subject to EIA (revised draft for comment)*, available at: www.china-eia.com/xwzx/11301.htm (in Chinese) (accessed 3 August 2017).

Ministry of Environmental Protection (2015), *Catalogue of Construction Projects Subject to EIA*, available at: www.mep.gov.cn/gkml/hbb/bl/201504/W020150420522354957849.pdf (in Chinese) (accessed 3 August 2017).

Modak, P. and Ginoya, N. (2013), "Challenges to integrate climate change considerations in environmental impact assessment" *33rd Annual Meeting of the International Association for Impact Assessment*, available at: www.iaia.org/conferences/iaia13/proceedings/Final%20papers%20review%20process%2013/Challenges%20to%20Integrate%20Climate%20Change%20Considerations%20in%20Environmental%20Impact%20Assessment.pdf?AspxAutoDetectCookieSupport=1 (accessed 3 August 2017).

National Development and Reform Commission (2013), *China's National Strategy for Adaptation to Climate Change*, available at: www.gov.cn/zwgk/2013-12/09/content_2544880.htm (in Chinese) (accessed 3 August 2017).

National Development and Reform Commission (2015), *China's INDC*, available at: www4.unfccc.int/submissions/INDC/Published%20Documents/China/1/China's%20INDC%20-%20on%2030%20June%202015.pdf (accessed 3 August 2017).

Oates, N., Conway, D. and Calow, R. (2011), *The "Mainstreaming" Approach to Climate Change Adaptation: Insights from Ethiopia's Water Sector*, Overseas Development Institute, London, available at: www.odi.org/sites/odi.org.uk/files/odi-assets/publications-opinion-files/7056.pdf (accessed 3 August 2017).

Rothman, D.S. and Robinson, J.B. (1997), "Growing pains: a conceptual framework for considering integrated assessments", *Environmental Monitoring and Assessment*, Vol. 46 Nos 1/2, pp. 23-43.

Ruhl, J.B. (2010), "Climate change adaptation and the structural transformation of environmental law", *Environmental Law*, Vol. 40, pp. 420-431.

Sadeleer, N. (2002), *Environmental Principles: From Political Slogans to Legal Rules*, Oxford University Press.

Shackley, S. and Wynne, B. (1995), "Integrating knowledges for climate change: pyramids, nets and uncertainties", *Global Environmental Change*, Vol. 5 No. 2, pp. 113-126.

Shear, M.D. (2017), "Trump will withdraw US from Paris climate agreement", *The New York Times*, available at: www.nytimes.com/2017/06/01/climate/trump-paris-climate-agreement.html (accessed 3 August 2017).

Wang, J. (2014), *Environmental Law*, Peking University Press (in Chinese).

Wang, X. and Gao, Q. (2015), "Policy and legal responses to climate change adaptation in China: new developments, new challenges", in Koh, K., Kelman, I., Kibugi, R. and Eisma Osorio, R.-L. (Eds), *Adaptation to Climate Change*, World Scientific, pp. 119-142.

Wende, W., Bond, A., Bobylev, N. and Stratmann, L. (2012), "Climate change mitigation and adaptation in strategic environmental assessment", *Environmental Impact Assessment Review*, Vol. 32 No. 1, pp. 88-93.

Wentz, J.A. (2015), "Draft NEPA guidance requires agencies to consider both GHG emissions and the impacts of climate change on proposed actions", *Environmental Law in New York*, Vol. 26 No. 4, pp. 57-63.

Further reading

State Council Information Office (2011), *China's Policies and Actions for Addressing Climate Change*, State Council Information Office, Beijing, available at: http://english.gov.cn/archive/white_paper/2014/09/09/content_281474986284685.htm (accessed 3 August 2017).

About the author

Qi Gao obtained her PhD degree from the School of Law, Western Sydney University. She lectures in Environmental Law at KoGuan Law School, Shanghai Jiao Tong University. Her research interests have been on environmental concerns in Asia Pacific regions, especially with regard to transboundary water resources management and climate change adaptation. Her doctoral thesis focused on transboundary legal disputes among riparian states of the Mekong River, which was published by Brill/Martinus Nijhoff in 2014. In addition, she has been doing research on environmental procedural rights, including access to information, public participation and access to justice. A more transparent, participatory and accountable decision-making process is key to environmental good governance in China. Qi Gao can be contacted at: gaoqi0304@gmail.com

5

Pro-poor adaptation for the urban extreme poor in the context of climate change

Md. Zakir Hossain and Md. Ashiq Ur Rahman

Urban and Rural Planning Discipline, Khulna University, Khulna, Bangladesh

Abstract

Purpose – The purpose of this paper is to examine pro-poor urban asset adaptation to climate variability and change. It constructs a conceptual framework that explores the appropriate asset adaptation strategies for extreme poor households as well as the process of supporting these households and groups in accumulating these assets.

Design/methodology/approach – Qualitative data are obtained from life histories, key informant interviews (KIIs) and focus-group discussions (FGDs). These data are collected, coded and themed.

Findings – This research identifies that households among the urban extreme poor do their best to adapt to perceived climate changes; however, in the absence of savings, and access to credit and insurance, they are forced to adopt adverse coping strategies. Individual adaptation practices yield minimal results and are short lived and even harmful because the urban extreme poor are excluded from formal policies and institutions as they lack formal rights and entitlements. For the poorest, the process of facilitating and maintaining patron–client relationships is a central coping strategy. Social policy approaches are found to be effective in facilitating asset adaptation for the urban extreme poor because they contribute to greater resilience to climate change.

Originality/value – This study analyses the empirical evidence through the lens of a pro-poor asset-adaptation framework. It shows that the asset-transfer approach is an effective in building household-adaptation strategies. Equally important is the capacity to participate in and influence the institutions from which these people have previously been excluded.

Keywords Bangladesh, Adaptation, Climate change, Asset, Urban poverty

1. Introduction

There is an emerging consensus that climate-change adaptations that are pro-poor can be achieved from ongoing development interventions (ActionAid, 2006; Agrawala, 2005; Huq *et al.*, 2007a, 2007b; Mitchell and Tanner, 2006). Huq *et al.* (2007b) indicate that to reduce climate-change vulnerability, urban adaptation strategies should target the most vulnerable citizens, i.e. those who are often associated with the poorest and most excluded groups. Scholars argue that improved urban development planning and the provision of public services and safe infrastructure are crucial for both the development and promotion of resilient cities (ActionAid, 2006; Stern, 2007; Tanner *et al.*, 2009). In addition to these planned adaptation measures, governments, donors and civil society actors must consider how they can support autonomous adaptation efforts by poor urban populations.

A robust social policy response is one that is rooted in an understanding of the risks the poor and extreme poor face in relation to climate change and climate vulnerability, and how to help them overcome this (Heltberg *et al.*, 2009, 2010). This research approaches this topic from the point of view of two impoverished communities in Bangladesh. It points to social approaches to assisting poor urban communities manage both climate change and climate volatility and support them as active agents in creating resilience through different household- and community-level interventions. These programmes include programmes social funds for community-based adaptations and asset-transfer. It identifies the processes through which the extremely poor can accumulate these assets and build resilience to the impacts of climate change. Further, it identifies how the extreme poor households' assets can be protected, and it explores how long-term resilience for the urban extreme poor can be built. This study may offer valuable lessons for policy adaptation that is pro-poor, as this concept is not yet widely appreciated in the climate-change-policy literature.

This research is set up as follows. Section 2 presents the conceptual framework of pro-poor asset adaptation and Section 3 defines the research strategy. Later, the results and discussion section identifies urban extreme poor's autonomous adaptation, role of formal institutions for building pro-poor asset adaptation, integration of pro-poor adaptation in urban planning, role of informal institutions for building pro-poor asset adaptation and role of social sector approaches for pro-poor asset adaptation.

2. Conceptual framework of asset adaptation of the extremely poor
Adaptation to climate change at the local level is largely made up of individual choices; in an urban context, however, collective action taken at the community and municipal levels is the most appropriate response for adaptation (Adger, 2005; Moser, 2010; Tanner *et al.*, 2009). Considering this, the conceptual framework of pro-poor asset adaptation to climate change identifies the process of building micro-level (household), meso-level (community) and macro-level (municipal) adaptation strategies that enable the urban extreme poor to protect themselves, or to recover from, the negative effects of the slow insidious changes in weather events whose patterns and intensity are associated with climate change.

Poor individuals or households take different physical, social, economic and political measures to reduce risks from extreme weather events, such as flooding, landslides or extreme temperatures. Assets (human, financial and social) available to poor households help them build their own individual adaptive practices (Moser, 2010). However, the adaptive responses of the individuals or households that are counted among the urban extreme poor are mostly short-term, individual and *ad hoc* impact-reducing efforts to save lives or protect property. Urban policies and institutions, both formal and informal, often place significant limitations on the urban extreme poor in their efforts to extend their individual strategies for long-term resilience to climate change (Devas, 2001; Huq *et al.*, 2007a, 2007b; Tanner *et al.*, 2009). Therefore, this sector of society needs asset-based planned adaptations that are more poverty focused so as to help build both their short- and long-term resilience (Heltberg *et al.*, 2009; Vernon, 2008).

At the micro level, the household should be considered the most important unit for the accumulation and management of resources. An effective approach for the extreme poor would be one that requires multiple entry points, and instruments that are protective (cash grants and/or food aid), preventive (basic services and infrastructures) and promotive (productive asset transfer, enterprise development, trainings and credits), that can create these entry points for the poor so they can build their household assets (such as financial, human, physical, livelihood-productive assets and household belongings) when all of these measures work together at least some of the time (Hulme and Moore, 2008). Households' access to productive assets, such as micro-enterprises, can reduce poor households' overdependence on one income source – their own labour. Preventive measures (such as the provision of basic services and housing improvements) can save both their human capital and their financial assets from sudden erosion (Devereux and Sabates-Wheeler, 2004). A combination of protective, preventive and promotive measures can enhance the resilience of households of extreme poor by developing new livelihood activities and making multiple assets available to them.

Assets available to households of the extreme poor help them buy their way out of risks – for instance, by being able to buy, build or rent homes that can withstand extreme weather in locations that are less at risk from flooding. Assets will also enable them to afford measures that help them cope with illness or injury or damage to their assets due to climate change. Given the insecurity of urban livelihoods, the asset-accumulation process is not necessarily sustainable for these households unless these new assets are also structurally protected (Banks, 2012; Hossain and Matin, 2004). Therefore, micro-level asset-adaptation strategies need to be integrated into meso- and macro-level strategies to protect households from asset erosion and to maximize the linkages between different inter-dependent assets.

At the meso level, urban communities that exist in an informal settlement also deploy different physical (e.g. clearing blocked drainage channels), socio-economic (e.g. community-based savings groups) and political measures (e.g. development of networks with political parties and NGOs for formal assistance) to withstand weather-related risks that arise from climate change (Wamsler, 2007; Wamsler and Brink, 2014). These are constrained, however, often by a lack of secure tenure and communities' physical and financial capacities to undertake large infrastructural projects. For example, developing a drainage system that actually stops or greatly reduces flooding – especially in high-density settlements on high-risk sites with little or no drainage infrastructure and space for new infrastructure – is usually beyond the means of community organizations. This is not to say that it cannot be done. Community-directed slum and squatter upgrading has achieved this, but this is where support from social funds has been provided and community-driven development has occurred.

Through social funds, the international community could channel external finance to small-scale community-based adaptation projects in a large number of communities of informal settlements (Heltberg et al., 2009). To protect households' assets and help them build greater resilience, households of the extreme poor need to connect with community-based adaptations. New livelihood activities can provide these households with the type of flexibility and wellbeing that can give them the confidence to participate in community-based risk-management strategies. This can facilitate the development of collective assets (such as saving groups, peer support groups and cooperatives) that can have positive influences on coping with and adapting to a wide range of stresses or shocks, including those arising from extreme weather. Human-capital-development services combined with learning-by-doing project activities can build organizational capacity into a collective entity for groups of the extreme poor. While the extreme poor lack access to important information related to risks and risk management, their active engagement in community networks can improve their access to the information they need to manage their risks.

The possibility for building resilience to climate change is much greater when community-based organizations build partnerships with local or national governments. In many countries, there are now national federations of slum and shack dwellers that have entered into partnerships with local governments and aid organizations. These partnerships have implemented programmes to address housing, upgrade slums and make improvements to service delivery and basic infrastructure in slums. However, it is also necessary to make changes in formal urban institutions and policies (macro level) that aim to help poor households and communities become more proactive in building asset adaptation to promote long term resilience in the face of climate change. For example, the local governments in Ho Chi Minh City and Cape Town called for resettlement of communities that were situated in the most risk-prone areas, in addition to ensuring improved construction and building regulations for the construction of low-income and informal housing (Baker, 2012). Of the more proactive approaches, both these municipal governments encouraged the adoption of new methods of construction that will enable communities to cope with anticipated flooding, such as elevating buildings and creating floating communities.

Social protection for vulnerable and marginalized groups (e.g. cash transfers) can enhance their preparedness for recovery after a climatic shock. At the macro level, institutional actors should undertake such actions as identifying adaptation options for the poor and marginalized that consider existing housing and social-sector policies that complies with the kind of asset adaptation vulnerable households and communities need. For example, the national government should give power and resources to local governments so they can help poor households and communities obtain safe, legal land sites for housing; the urban poor should also have improved access to justice.

This framework will be used to understand adaptive responses of the extreme poor to the extreme weather events that arise from climate change. It will explore how social sector programmes help the extreme poor build asset adaptation for short- and long-term resilience. It will also identify how urban institutions and policies constrain their adaptation strategies in relation to the poor and identify the gaps in pro-poor asset adaptation for the urban extreme poor in Bangladesh (Figure 1).

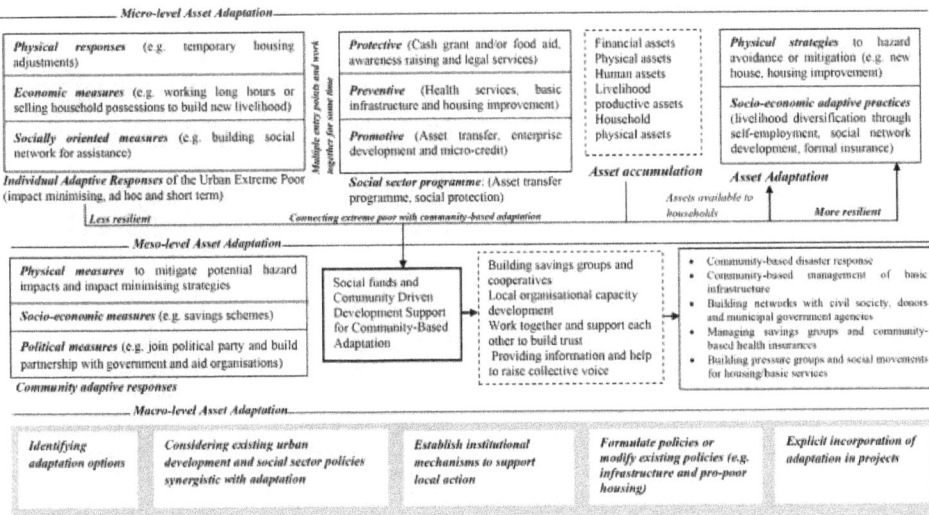

Figure 1. Pro-poor asset adaptation framework for the urban extreme poor

3. Materials and methods

This paper is based on a qualitative research design process. At the micro level, the empirical research consisted of case studies of settlements in Bangladesh. Slums were selected for review for which sufficient information already existed in the literature. The settlements of Karail and Kamrangirchar were selected because of the existence of project-based interventions for households of the extreme poor in these communities, with different approaches being available for review. In Karail, a community-based approach is being implemented that includes both poor and extreme poor households; in Kamrangirchar, extreme poor households are supported through an asset-transfer programme (Figure 2).

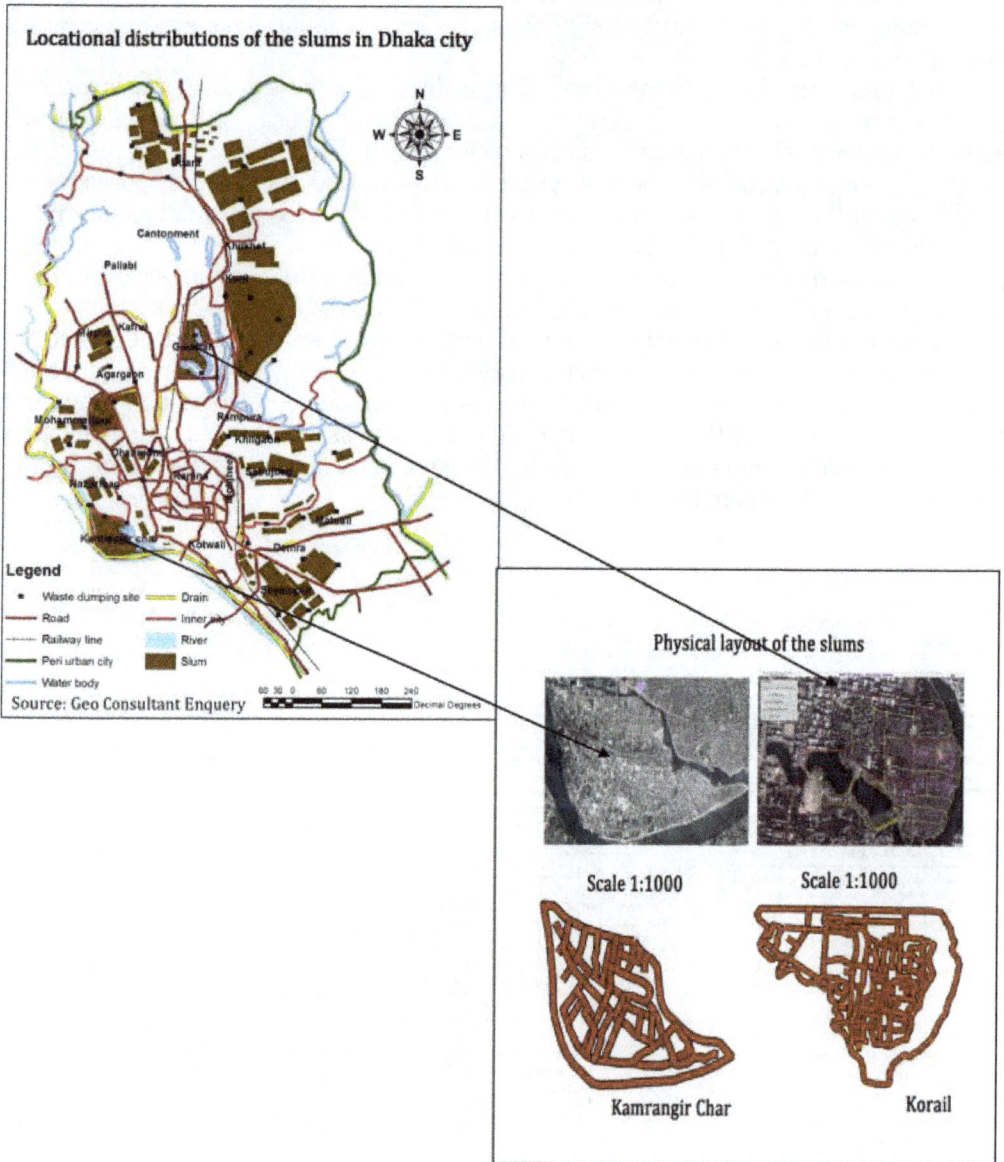

Figure 2. Map of case study settlements

The data collection technique included interviews on the life histories of extreme poor households. Forty interviewees (20 respondents each from Karail and Kamrangirchar) were selected for face-to-face life-history interviews through a technique wherein local community leaders recommended which households were the most suitable candidates for this study. The interviewees were selected from both socially vulnerable (old, widow/divorced and chronically ill/disabled) and professionally vulnerable groups (rickshaw pullers, maids and beggars). These life-history interviews were supplemented by surveys of key informants, and also focus-group discussions and the analyses of grey materials. For a broader testing of the findings, the researcher held dialogues with representatives of public-service-delivery organizations, members from community-based organizations and the academic community, policy makers, civil society organizations and knowledgeable members of the two selected settlements. In addition, focus-group discussions were used to identify the relevant urban institutions and processes at the two settlements in the case study. Focus-group discussions were also undertaken to validate the information.

The qualitative data that were gathered from different interviews were organized under a number of themes. The researcher read through the transcripts and noted down themes as they appeared in the interviews. For example, if the interviewees talked about their different aspects of poverty, such as food insecurity, housing, water supply and sanitation, shocks and stresses, the researcher noted this down as "multidimensional aspects of poverty". At the bottom of each page, the researcher then noted down the main themes, such as the multidimensionality of poverty, vulnerability, urban institutions and the role of programmes. From this initial detailed analysis, the researcher reviewed the themes that had been pulled out of the interviews and grouped these into wider thematic categories, such as service accessibility, migration, labour market participation and barriers to entry into the labour market. Finally, the researcher used electronic copies of his transcripts to piece together the segments of data that represented each theme. Following this, the researcher developed qualitative analysis through analysing in detail what the extreme poor said about these themes and what they might signify in their everyday practices and how they negotiated within their everyday lives.

4. Results and discussion

4.1 Urban extreme poor's autonomous adaptation

The study data from the household interviews reveal that households of the urban extreme poor employ a range of measures to reduce or adapt to their level of disaster risk shortly before they incur the impact of the potential hazard. Moreover, during and following the impact, they also deploy *ad hoc* response-and-recovery measures. The range and diversity of individual adaptive practices can therefore mainly be categorized according to two objectives:

(1)   impact-minimizing strategies; and

(2)   preparedness measures for response and recovery.

Most of the impact-minimizing actions have become integral parts of coping and adapting practices that had been generated in past experiences. For example, in Karail, the extreme poor construct barriers at their front doors (5 households out of 20), increase furniture height (6 households) and build higher plinths (12 households). In the settlement of Kamrangirchar, similar physical coping-and-adaptive strategies are also found in the urban periphery. This settlement's extreme poor households also deploy impact-minimizing strategies, such as placing barriers across their front doors (4 houses out of 20), putting furniture onto bricks (3 houses), arranging high storage facilities (3 houses) and building high plinths (14 houses).

The study found some physical measures taken by the urban poor, such as building two-storey structures and using combinations of construction materials to build residential units that have the potential to reduce the risk of flooding. Only 4 of the total of 40 extremely poor households were found to be living in the two-storey houses. To help adapt to very high temperatures, creepers were grown in courtyards so as to cover roofs, and other materials were put on roofs to also reduce the heat (3 houses in Karail and 4 in Kamrangirchar); a number of households (12 households out of 40) used some form of false ceiling or canopy made out of cloth (a popular practice in rural areas, adopted in urban houses). Generally, closely spaced structures create shaded courtyards that are used as open space for ventilation. Household activities are often held outdoors during frequent power shortages. The use of different insulating materials reduces the heat that comes from corrugated sheet iron roofing and partitions.

The urban extreme poor also use specific non-hazard political as well as economic- and social-oriented measures to reduce or mitigate the increased risk levels that are associated with climate variability and change. Further, respondents from male-headed households reported that engagement in two jobs, simultaneously, limits the number of working hours in labour-intensive jobs; this is one strategy through which household heads try to reduce or overcome the impacts of hot and humid weather, during summer. This research found that two male breadwinners in Karail maximize their hours worked by pulling a rickshaw in the morning and running a street-peddling business in the afternoon. The elderly heads used this strategy, during the summer months, to cope with the exhaustion caused by undertaking physical labour in the extreme heat. In Kamrangirchar, two female heads (out of 20) used this strategy to get relief from heat stress, as both the weather and the associated power cuts make it impossible to work the whole day. Respondents from female-headed households reported that they worked as maids in the morning and engaged in the shoe-making industry in the afternoon or evening.

The urban extreme poor of both settlements employ a number of socially oriented measures. They practice highly effective forms of reciprocity and attachment with neighbours, friends and family members who offer them assistance at times of adversity. The common mechanisms these extreme poor families use to cope with seasonality are borrowing money and/or goods. Male respondents explained that families are always very cautious about their repayment of a loan. Repayment is considered very important for obtaining another loan the next time they face adversity. Common strategies used to repay these loans are either saving money by going hungry or working long hours when the rain stops. These financial coping strategies can create further obstacles (such as the erosion of physical strength) for these households. Other measures relate to education and include sending children to study outside of the slum area. In Karail, one female household head sent her daughters to an NGO school, outside the slum area, where they are offered not only free secondary school education but also free daily meals. In Kamrangirchar, two female heads (out of 20) sent their children to their parent's houses, which are in a rural area, as they can get free education and female stipends from government schools. This stipend means a monthly allowance to support their need for buying educational materials.

In 2012, the community network also played a dominant role in tackling the eviction threats of the Karail bustee. The community committees mobilized the entire community to demonstrate against threats of eviction. On 4 April 2012, they put pressure on those at higher political levels, e.g. ministers, their member of parliament, the mayor, the ward councillor and public agencies that were involved in the evictions in the Karail slum, to put a stop to this unethical practice. In addition, the community leaders developed a network with members of civil society and legal aid institutions to initiate a legal action against the evictions. They filed a writ petition in the High Court to suspend the evictions from the Karail slum, until further orders. During the interview process, a male community leader reported that the leaders were also actively involved in mass campaigning against the evictions of the Karail squatter settlement. They did so by going through the national newspapers and print media to seek the support of the people of Dhaka City. This was a success story as these community leaders received financing and relief, and protests and dialogues with the members of formal political systems (e.g. ministers, political leaders). Following their writ petition, the High Court issued a stay order on evictions from the Karail slum. In Kamrangirchar, however, the urban extreme poor residents did not report any politically oriented measures apart from the rickshaw pullers' participation in the local rickshaw pullers' association.

The urban extreme poor of both settlements coped with or adapted to extreme-flooding events by taking a range of preparedness measures for response and recovery. Most measures were taken shortly before the occurrence of impacts from the potential hazard. In Kamrangirchar, most of the urban inhabitants moved to the nearest safe places, such as primary schools, embankments, rail lines, highways and government offices. A female landlord in Karail explained that during the flooding of 1998, moving to safer areas is not a preferred option; moving means losing assets and possibly losing the right to live in the area. During an emergency, the most practised option is to sleep on furniture above the flood level and use movable cookers for food preparation. Some households shared the services of their unaffected neighbours. The community leader also reported some temporary measures to reduce the impacts of flooding and water logging, such as building high barriers at the front doors, creating outlets from the house for easy flow of water, building high stilts inside the rooms and community initiatives to clean drainage and move sick children, the disabled and the elderly to hospitals and safer spaces within the neighbourhood.

The interviews with the extreme poor who faced the flooding event of 1998 revealed that they used their savings to address food storages and their family's medical needs. When the floodwaters did not recede, during that event, households borrowed money to meet food costs. The households of the urban extreme poor of both settlements also deployed some preparedness recovery strategies to help them quickly recover from the impact of a hazard/disaster (e.g. damage to housing and loss of income), and bounce back to their former condition. From the interviews conducted with the extreme poor, this research found that there was a collective effort between home-owners and tenants in rebuilding houses and improving living conditions. Though most of the extreme poor stayed in rental accommodations, they were actively involved in the post-disaster-reconstruction process that took place within their housing cluster, including cleaning the courtyard, rebuilding community kitchens, repairing floors, increasing plinth levels and exchanging materials for plinths, roofing and walls. They took out loans and got help from household members or neighbours for this purpose. Home owners revealed that they obtained loans from their community saving groups to pay for repairs. The residents of both settlements used some common strategies to address their lack of income, food insecurity and health problems, such as using savings to build alternative livelihoods; working long hours and engaging in two jobs, simultaneously; selling some household possessions; taking loans from informal sources; begging; and borrowing money from an employer.

4.2 Role of formal institutions for building pro-poor asset adaptation
In Dhaka City, the institutional arrangements related to urban planning, basic services and infrastructure are complex. They include housing, water supply and sanitation, roads and drainage. Services are delivered by a mix of central, urban and local governance agencies

that have limited resources, weak administrative capacity and little coordination. Dhaka City Corporation (DCC) is the city's urban political administration; its specialized agencies include the Capital Development Authority-RAJUK; Dhaka Electric Supply Authority (DESA); Dhaka Metropolitan Police (DMP); Titas Gas; central government institutions, such as those responsible for land administration, public works, education and health; the Bangladesh Telephone and Telegraph Board (BTTB); the Bangladesh Road Transport Corporation (BRTC); and the Dhaka Water and Sewerage Authority (D-WASA). Figure 1 shows the urban institutional arrangements and their relationships. As explained above, national policies for housing and basic services do not provide much legislative support and protection to the urban poor (IGS, 2012). In the absence of legislative support, the government agencies are not willing to assist the urban poor in accessing housing and basic services in Dhaka City's urban slums (Figure 3).

There is a lack of clarity regarding the relative roles of the main public institutions that are tasked with intervening in the housing sector (World Bank, 2007). These institutions include the capital development agency, RAJUK; the National Housing Authority (NHA); the Urban Development Directorate; and the Public Works Department of the Ministry of Housing. The NHA has jurisdiction over all of Bangladesh and is responsible for producing and implementing the National Housing Policy. However, it does not have the means to enforce its policies. RAJUK's jurisdiction covers the Dhaka Metropolitan Area, including the city proper. However, RAJUK's mission has never been to provide housing for the poor. Nevertheless, it has been known to provide serviced lots of a minimal size (100 square meters). But, given land prices in the city, these lots are totally unaffordable for poor households. Dhaka City Corporation and special purpose agencies, such as DWASA, are responsible for delivering basic services, such as water supply and sanitation, roads, drainages and solid waste management. Neither of these urban government institutions can adequately perform their functions due to the severe resource constraints they face and their limited authority. The larger squatter settlements' lack of legal tenure further complicates service delivery, as neither DWASA nor DCC are willing to invest in settlements such as Karail.

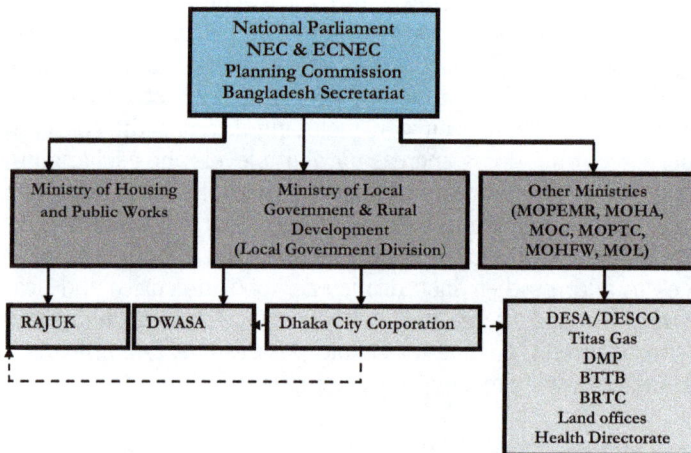

Figure 3. Mapping of urban public institutions in Dhaka CITY

4.3 Integrating pro-poor adaptation in urban planning

As a part of disaster preparedness, the Dhaka Metropolitan Development Plan (DMDP) set out the standard for land-use and floodplain management as per the guidelines of the National Flood Action Plan. This plan recommends that at least 12.1 per cent of the low-lying areas be preserved as retention ponds for flood water (DMDP, 1995). The DMDP's structure plan designates two categories of flood flow zones: the main flood flow and the sub-flood flow zones, which are based on the two flow directions of the rivers that adjoin the city and are the main tributaries in Dhaka. According to the DMDP (1995), it is strictly forbidden to develop land for residential, commercial and industrial use, including raising the land level with fill. The only permitted land uses include agriculture; dry-season recreation facilities; ferry terminals; and excavation of mineral deposits (e.g. dry-season brick works) that will not cause adverse hydraulic effects. In the sub-flood flow zones, development that is compatible with the rural nature of the area will be permitted insofar as it does not to disturb the flood flow.

The Capital Development Authority, RAJUK, is the public institution responsible for implementing the recommendations of the DMDP; it has the following legal tools to control land use and city growth: the DMDP 1995, comprising the Urban Area Plan and Detailed Area Plan (structure plan); the Town Improvement Act, 1953; Building Construction Rules, 1996; Private Residential Land Development Rules, 2004; The Natural Water Body, Open Space, Park/Play Ground Preservation Rule, 2000; and the recently approved *Mohanagar Imarat Nirman Bidhimala*, 2006 (Building Construction Rules, 2006). These laws and rules have several weaknesses that affect their ability to control unplanned growth in the flood flow zones and provide incentives for the urban poor to build preventive measures against extreme weather events.

4.4 The role of informal institutions for building pro-poor asset adaptation

Amid the lack of formal urban institutional support, informal institutions (such as, local political elites and community-based organizations or CBOs) emerge through which the urban poor are able to access housing and basic services. This research has found a range of informal institutions, including externally connected political leaders, internally connected local elites, CBOs, local credit groups and organizations for the urban poor. These organizations provide support for communities in Karail to build autonomous adaptations to climate change. In addition to these informal institutions, formal political systems, NGOs, aid organizations and legal-aid civil-society organizations also play an important role in building individual and communal adaptive practices. The informal institutions interact with public (e.g. Dhaka City Corporation, DWASA) and private agencies (e.g. NGOs and legal-aid organizations), and attempt to influence the provision of assets so as to enable these communities to help those in vulnerable situations adapt to the potential hazardous conditions that are related to climate change.

This research has investigated the role of informal institutions in building or constraining the urban poor's adaptation practices in the settlements covered in the case study. The above-mentioned institutions have been mapped out according to the level of interaction of the urban poor with these institutions in terms of accessing assets and services for adaption (Tables I and II).

The participatory institutional mapping, of both of the settlements in this study, found patron-client relationships, at multiple levels, that provide varying degrees of support. While the poor and extreme poor households of both settlements lack a formal set of entitlements, they try to manage uncertainty and improve access to employment, finance and physical and social support by maintaining patron-client relationships. In this context,

Table I. The role of informal institutions for adaptation practices in Karail

Institutions available to poor urban households	Access to assets	Asset adaptation
Internally connected leaders and economic elites within the settlement	Employment, financial assets and emergency credit	Household economic measures (e.g. regular employment, employment diversification). The emergency credit helps households to use physical measures
Externally connected leaders build informal relation with public service delivering organizations	Access to water supply	Households can access water for their daily use within their housing compound, which saves time for income generation. But water is insufficient for meeting daily needs. During monsoon, water often gets contaminated due to unhygienic pollution entering pipes at weak connection points. Moreover, poor households use ring wells to collect water from this unhygienic pipe system, which increases health risks. When flood strikes, households have to collect water from alternate source (tube well) from further distance, at a higher price and with longer waiting time due to increased people using the same water source
	Access to electricity	Access to electricity connection provides an opportunity for poor households to increase their power usage to reduce heat stress during summer. But in Karail during hot weather, the increased use of informal power frequently causes electrical connections to short-circuit, resulting in fires (which can spread quickly in dense settlements)
Civil society organizations and community groups for tenure security	Tenure security	Households can deploy preventive and impact minimizing physical measures to climate variability
NGOs and community groups	Accessing services (e.g. credit, health services, education and emergency relief)	Household can deploy various economic and socially oriented measures to reduce the impacts of climate stresses and shocks
Formal political systems through externally connected leaders	Access to emergency relief	When disaster strikes, households get access to formal support (e.g. relief and financial grants)
Elite neighbours near the Karail settlements	Employment, financial credit and emergency relief	It facilitates household economic measures; and emergency relief helps to build households' emergency preparedness
Locally formed groups (e.g. Bazaar (market) committee, regional committee, youth groups and women's groups)	Emergency services	Move vulnerable individuals, households to safer places, distribution of flood warning system and community collective actions for preparedness recovery

Source: This is based on the key informant interviews undertaken at Karail, involving three community leaders; the findings of key informants are validated by one focus group discussion. The group consisted of two house/room owners and five tenants

Table II. The role of informal institutions for adaptation in Kamrangirchar

Institutions available to poor urban households	Access to assets	Asset adaptation
Strong relation with house owners	Tenure security, services, emergency food and credit	Households can take physical and economic measures to reduce the impacts of extreme climate events
Patrons outside the community (e.g. employer)	Employment, credit and food support	Households can take impact minimizing non-hazard measures
NGOs' credit services outside the community	Access to financial assets	Households can diversify their income and employment; and access to emergency credit for preparedness recovery
NGO water supply and sanitation, environmental health services in some slum clusters	Accessing services	Human capital development; and community-based actions to improve basic services

Source: This is based on participatory methods including two key informant interviews and one focus group discussion undertaken at Kamrangirchar, involving two non-poor house owners and five extreme poor tenants

Jahan *et al.* (2011) argue that the ability to build and sustain multi-level social networks acts as the main foundation for survival and improvement in the informal settlements of Dhaka City.

The household interviews revealed that the urban poor do not just have one patron. It was found that households of the urban poor seek solutions to their multiple needs, at different levels, including from employers, landlords, local political leaders, local representatives, community leaders and moneylenders. Households of the extreme poor further seek to improve and extend their social contacts by seeking to establish and maintain a patron by being trustworthy, hardworking and reliable. This approach may offer future returns in terms of borrowing arrangements for cash or food. For example, a good working relationship with a garage owner can lead to short-term loans on reasonable terms. Those who work as maidservants consider their employer as one of their important social contracts. In a sudden crisis, these contacts sometimes give financial help and suggestions. However, household heads who experience a lack of physical vigour or ill health find that this constrains their ability to actively construct or benefit from social capital through association. For example, it was found that garage owners are reluctant to offer employment to the elderly who wish to pull rickshaws; employers and moneylenders often refuse to provide credit to households where breadwinners are sick or disabled.

Savings schemes can be drawn on to help cope with a wide range of stresses or shocks, including those arising from extreme weather. The institutional mapping (as shown in Tables I and II) reveals that in both Karail and Kamrangirchar, NGOs and/or cooperatives operate formal credit-and-savings schemes that offer loans to the urban poor for building income-generating activities. They also provide credit to help the urban poor adapt to stresses related to climate. Additionally, these schemes also help people use different physical and economic recovery-preparedness measures. In formal credit arrangements, the urban poor must participate in compulsory savings schemes for 6 to 12 months to get access to credit. Loan repayment can guarantee further access to loans. Banks (2012) argues that compulsory savings periods often limit the role of formal loans as a source of emergency financing for the urban poor. However, the research identifies that, in both settlements in the case study, formal credit systems operated by NGOs and CBOs rarely select extreme poor households as beneficiaries. This is partly because of self-exclusion. That is, the extreme poor households reported that they do not have the ability to take on risk, which is key to becoming successful entrepreneurs. So, initially, these households do not want microcredit and, hence, they do not want to join in the community group because they feel that they would be unable to pay back the loan and would therefore be stuck with debt for which they would eventually be forced to sell off their few possessions.

There is also an element of social exclusion since, for both economic and social reasons, members in the credit-and-savings schemes do not want to associate with members of extreme poor households. Community members refer to households that are in extreme poverty as the beggar class and as poor helpless households. The latter either have only one adult income earner, who is often a woman, or the adult members of these households are not in a position to use their existing human capital for various reasons, such as childcare and sickness. The community members consider that these poor households have a low capacity to make regular savings and to invest the money they borrow. As the members of the credit-and-savings group must shoulder joint responsibility for repayment of the loan, they prefer members who are either moderately poor or non-poor slum dwellers. Hence, this excludes the extremely poor. This exclusion from access to financial services limits these people's capability to gain income-and-asset diversification and, consequently, their vulnerability remains unchanged.

4.5 Role of social sector approaches to pro-poor asset adaptation
This section examines evidence from two social sector projects: the Urban Partnership for Poverty Reduction (UPPR) project and the DSK-Shiree project, which are being implemented in Karail and Kamrangirchar, respectively. Table III shows how social-sector approaches facilitate asset adaptation or resilience strategies for the urban extreme poor in these two communities. The UPPR project focuses on a participatory approach to mobilizing communities. Here, it uses different structures and seeks to identify the problems different groups of the poor face; it then prioritizes actions to mitigate these problems. Although the project used a democratic process to identify community leaders, in reality, the existing leadership, local elites and their supporters hold the committee positions on these projects. Therefore, the UPPR project can be viewed as a project that serves to distribute patronage and not one that provides an enabling environment for the enhancement of the extreme poor. The Community Development Committees within the project appear to consider the extreme poor only in terms of making demonstrations of charity and public largesse. This means that these types of promotional measures suffer from tokenism, marked by a modest transfer of resources. As a consequence, the UPPR project has less impact on households of the extreme poor in terms of helping them develop asset-adaptation strategies. Despite this, the project has devised a number of collective physical-adaptation strategies for the slum dwellers in Karail. These have helped this community improve drainage systems and reduce waterlogging risks in the slum. The UPPR project's savings-and-credit scheme can be drawn on to help the poor and extreme poor adapt to shocks and stresses, including those arising from climate variability.

Despite the drawbacks, these community-based adaptation strategies contribute to the adaptive resilience of the extreme poor, including both individuals and households. The UPPR project intends to bring transformative changes, such as access to legal services or tenure security through building strategic partnerships between the city and municipal government and local communities. However, several institutional barriers may constrain the sustainability or long-term support of municipal governments in Dhaka City, including inadequate laws and rules and financial capacity, a lack of incentives for the urban poor and

Table III. The role of social sector approaches for pro-poor asset adaptation

Level of interventions	Karail	Kamrangirchar
Household level asset-based adaptations through the UPPR and DSK-Shiree project	Self-employment for some extreme poor individuals who were unemployed and wanted to change job Additional labour mobilization for few extreme poor households Informal support networks Use of streets as rent free business yards	Economic diversification through self-employment Skilled employments for adult children Additional labour mobilization for household economic diversification Adjustments in small business enterprises (e.g. limited stock, physical adjustments, covering stock with plastics, arranging higher storages for the stocks) Move to safer places Accumulation of physical assets Sending small children to school Home-based enterprises Improvements in food intake and health seeking behaviour
Community-based adaptation through the UPPR and DSK-Shiree project	Community mobilization and representative community organization Building community infrastructure to prevent or avoid threats Community-based savings and credit schemes	Social organization and capacity building Development of climate proof community-based water supply and sanitation services Collective savings and business enterprises
Formal institutional responses through the UPPR and DSK-Shiree project	Dhaka City Corporation work with the CDCs in Karail to provide infrastructure and services D-WASA decided to provide legal water supply in the slum	No support from city and municipal organizations

Source: Findings summarized from 40 extreme poor interviews

a fragmented approach to service delivery. The UPPR project used community mobilization to identify and implement their own development priorities, but these actions have had fewer benefits of consequence for extreme poor households. As structural inequalities underlie the community structures of the informal settlements in Dhaka City, the committees within the UPPR project reinforce, rather than break, these pre-existing inequalities. Overall, the local elites appear to have more power and resources, while the poorest remain as dependent as ever on the local elites for gaining access to the resources the UPPR project provides.

The DSK-Shiree project's asset-transfer approach has been more effective in building self-help strategies, as compared to the UPPR project. This asset-transfer project takes a wider approach to social protection. It recognizes that social protection must address not just the economic risks associated with poverty but also the adverse structural context that extremely poor households face. Therefore, the DSK-Shiree focuses on providing multiple assets and also on the interconnectedness of these assets. This study suggests that the protective and preventive instruments offered should give extreme poor households the security they need to invest their time and effort in the "promotional" components of resilience (e.g. turning a small micro-enterprise's asset transfer into a viable income-generator). It is also found that this project's asset-accumulation strategies positively facilitate asset adaptations for extreme poor households. Economic diversification through self-employment, additional labour mobilization and skill employment is an effective practice that not only helps extreme poor households be less affected in the case of future hazards/disasters but also enables them to recover more quickly from the impacts of these hazards.

Extremely poor residents in Kamrangirchar are able to deploy hazard-avoidance measures, such as moving their small enterprises to a safer house and location. They also use some additional preventive measures for small businesses, such as raising plinths, arranging higher storage facilities, carrying limited stock and running a seasonal business so as to avoid the negative effects of weather hazards. Small, home-based enterprises are important for households with small children. This type of business allows household members combine income earning with caring for children. These families' women also have enough time to cook and serve family members fresh food, which contributes to greater resilience to extreme temperatures. Improvements in income also help extreme poor households in Kamrangirchar earn money to pay for additional power usage, during summer.

The importance of formal education for children was also identified in Kamrangirchar. Here, it was revealed that children's educational levels have a positive effect on the extreme poor's risk-reducing strategies through family members' future formal employment; their awareness and understanding of current risk levels; and access to and provision of information on risk reduction. Aside from these asset adaptations, the project also aids in reducing extremely poor respondents' physical-adaptation deficit through the construction of 39 improved water points and 62 hygienic pit latrines. In addition, the project's savings-and-credit schemes contribute to greater resilience and reduced climate risks. The practice of regular savings has both instrumental benefits (the ability of savers to access funds when necessary) and organizational benefits (the relationships of trust built up within small savings groups that allow their members to work on collective solutions to larger problems).

4.6 Discussion

The pro-poor asset-adaptation conceptual framework and the analysis of autonomous and planned asset-based adaptation in the Bangladesh context provide some important insights:

- The extreme poor have already coped with extreme weather events through deploying different physical adaptations and also adjustments in human behaviour. However, the strategies they use to cope with extreme events are short lived, ad hoc and impact minimizing, and even adversely lead to asset-based poverty traps. Therefore, the extreme poor households in this study need a set of integrated interventions that work together for over the long term. This will help them build asset portfolios that can be a key determinant in helping them adapt to extreme weather events and climate variability.

- External supports must be available at the neighbourhood level to improve the capacity of community organizations to develop community-level infrastructure that prevents or mitigates the adverse effects of extreme climate events.

- Households' assets need to be structurally protected.

Without this, positive changes can quickly be reversed if households face a sudden shock. Hence, meso- (community-based initiatives) and macro-level interventions (such as, floodplain management, infrastructure development and the provision of basic services) are important to protect household assets from destruction and/or erosion.

This investigation found that, for the settlements in this case study, protective and promotive components of an asset-transfer programme can have a positive impact on building the asset-based strategies of extreme poor urban households. It also shows that cash transfers can do more than just allow people to cope with shocks, especially if the amounts transferred are large enough. In this context, Hulme and Moore (2008) argue that the poorest people cannot benefit from a single "magic bullet" (such as, microcredit, bed nets and women's groups). Rather, they need a carefully sequenced set of supports that provide livelihood security; confidence building and skill development; asset transfer; and support for and institutionalization of their improved position within the local economy and society (Hulme and Moore, 2008).

Where there are representative community-based organizations, the possibilities of building resilience to climate change are much greater (UN-HABITAT, 2011). In many countries, there are now national federations of slum and shack dwellers that have community-based savings groups as their foundation (CPRC, 2008; Dodman *et al.*, 2009; UN-HABITAT, 2011). Community-managed savings groups constitute a widespread mechanism that is used by the urban poor in the slums of Dhaka City (Jabeen *et al.*, 2012). Local governments and aid agencies can support such savings schemes by backing their development into larger networks of savers – thereby helping to further spread their risks – and engaging in the co-production of housing, services and infrastructure. The savings patterns among the inhabitants, in both of the settlements in the case study, exemplify such an opportunity. While municipal governments are best placed to address the adaptation needs of the urban poor, the central government is reluctant to hand over power and resources to a democratically elected government for fear of its capture by the opposition. This greatly constrains the powers, functions and financial capacities at the municipal level. Without a change in the formal governance structures, the urban poor will continue to find it difficult to get access to a formal set of entitlements, such as housing, basic services and social protection.

5. Conclusion

This research contributes to the existing asset-adaptation frameworks by lending to a greater understanding of the asset-adaptation strategies and processes of extreme poor households and groups. Existing asset-adaptation arguments point to the need of the poor in urban communities to cope with extreme weather events through a range of measures. Both formal and informal contexts can provide an enabling environment within which actors can operate to both protect and/or adapt their assets. For the extreme poor in urban areas, access to adaptation assets becomes limited when the adverse structural context that comes from formal and informal institutions and policies are not considered. This research addresses this gap by analysing individual adaptive responses and their limitations, using the case of Dhaka City in Bangladesh.

Extreme poverty is an extreme stage of economic, social and political deprivation that results from the continuous erosion of the human, financial and household physical and social capital, of the poor, that result from unforeseen events. At the micro-level, it is necessary to pay special attention to each of the issues the extreme poor face. In this context, this requires the application of significant additional resources to further expand the asset-transfer approaches that are already successfully working in the regions of Bangladesh to those facing the highest incidence of extreme poverty. Meso-level interventions, especially social funds and community-driven supports should be connected with micro-level interventions to facilitate the kind of social organizations that ensure that the mobilization of vulnerable and marginalized people becomes a vehicle for breaking, rather than reinforcing, existing structural inequalities.

A strong community-based organization can negotiate and build pressure on the formal governance structures so as to ensure access to a formal set of entitlements, such as housing (in the form of tenure security) and infrastructure. Household- and community-level adaptation strategies are constrained, however, by a lack of secure tenure and household capacity, with most communities having neither the physical nor financial capacities to undertake large infrastructure projects. Thus, macro-level interventions that include infrastructure works, basic services and tenure security can protect households and communities from asset erosion and also generate opportunities to further the kind of asset accumulation that will allow these vulnerable people to exercise resilience in the face of climate-change hazards.

References

ActionAid (2006), *Unjust Waters: Climate Change, Flooding and the Protection of Poor Urban Communities – Experiences from Six African Cities*, ActionAid, London.

Agrawala, S. (Ed.) (2005), *Bridge over Troubled Waters, Linking Climate Change and Development*, Organisation for Economic Cooperation and Development (OECD), Paris.

Adger, W.N. (2005), "Scales of governance and environmental justice for adaptation and mitigation of climate change", *Journal of International Development*, Vol. 13 No. 7, pp. 921-931.

Baker, L.J. (Ed.) (2012), *Climate Change, Disaster Risk and the Urban Poor: Cities Building Resilience for a Changing World*, The International Bank for Reconstruction and Development, Washington, DC.

Banks, N. (2012), "Urban poverty in Bangladesh: causes, consequences and coping strategies", BWPI Working Paper [172], Brooks World Poverty Institute, Manchester.

CPRC (2008), *The Chronic Poverty Report 2008-2009*, Chronic Poverty Research Centre, Manchester.

Devereux, S. and Sabates-Wheeler, R. (2004), "Transformative social protection", IDS Working Paper [232], Institute of Development Studies, Brighton.

Devas, N. (Ed.) (2001), *Urban Governance and Poverty: Lessons from a Study of Ten Cities in the South*, The School of Public Policy, University of Birmingham, Birmingham.

Dodman, D., Mitlin, D. and Rayos, C.C.J. (2009), "Victims to victors, disasters to opportunities: community driven responses to climate change", *Proceedings of Fifth Urban Research Symposium, Marseille*, 28-30 June, 2009.

Heltberg, R., Siegel, B.P. and Jorgensen, L.S. (2009), "Addressing human vulnerability to climate change: towards a no-regrets approach", *Global Environmental Change*, Vol. 19 No. 1, pp. 89-99.

Heltberg, R., Siegel, B.P. and Jorgensen, L.S. (2010), "Social policies for adaptation to climate change", in Mearns, R. and Norton, A. (Eds), *Social Dimensions of Climate Change*, World Bank, Washington, DC.

Hossain, N. and Matin, I. (2004), "Engaging elite support for the poorest? BRAC's experience with the ultrapoor programme", CFPR-TUP Programme Working Paper [3], BRAC Research and Evaluation Division, Dhaka.

Huq, S., Kovats, S., Reid, H. and Satterthwaite, D. (2007a), "Reducing risks to cities from disasters and climate change", *Environment and Urbanization*, Vol. 19 No. 1, pp. 39-64.

Huq, S., Reid, H., Lankao-Romero, P. and Satterthwaite, D. (2007b), *Building Climate Change Resilience in Urban Areas and Among Urban Populations in Low and Middle-Income Nations*, Draft paper prepared for Rockefeller Foundation.

Hulme, D. and Moore, K. (2008), "Assisting the poorest in Bangladesh: learning from BRAC's targeting the ultra poor programme", BWPI Working Paper [1], University of Manchester, Manchester.

IGS (2012), *State of Cities: Urban Governance in Dhaka*, Institute of Governance Studies, BRAC University, Dhaka.

Jabeen, H., Allen, A. and Johnson, C. (2012), "Chapter 30: built-in resilience: learning from grassroots coping strategies to climate variability", in Hoornweg, D., Freire, M., Lee, J.M., Bhada-Tata, P.

and Yuen, B. (Eds), *Cities and Climate Change: Responding to an Urgent Agenda*, World Bank, Washington, DC, Vol. 2, pp. 776-796.

Jahan, F., Hulme, D., Roy, M. and Shahan, A. (2011), "Re-framing the problem: from climate change in urban areas to urban governance in an era of climate change", Proceedings of 40 years of Bangladesh: Retrospect and Future Prospects Conference, *Institute of Governance Studies, BRAC University, Dhaka*, 26-28 November, 2011.

Mitchell, T. and Tanner, T.M. (2006), *Adapting to Climate Change: Challenges and Opportunities for the Development Community*, Tearfund, Teddington.

Moser, C. (2010), "A conceptual and operational framework for pro-poor asset adaptation to urban climate change", in Hoornweg, D., Freire, M., Lee, J.M., Bhada-Tata, P. and Yuen, B. (Eds), *Cities and Climate Change, Urban Development Series*, The World Bank, Washington, DC, pp. 225-253.

Stern, N. (2007), *The Economics of Climate Change: The Stern Review*, Cambridge University Press, Cambridge.

Tanner, T., Mitchell, T., Polack, E. and Guenther, B. (2009), "Urban governance for adaptation: assessing climate change resilience in ten Asian cities", IDS Working Paper [315], Institute of Development Studies, Brighton.

UN-HABITAT (2011), *Global Report on Human Settlement 2011*, Cities and climate change, Earthscan Publications, London.

Vernon, T. (2008), "The economic case for pro-poor adaptation: what do we know?", *IDS Bulletin*, Vol. 39 No. 4, pp. 32-41.

Wamsler, C. (2007), "Managing Urban Disaster Risk Analysis and Adaptation Frameworks for Integrated Settlement Development Programming for the Urban Poor", PhD Thesis, Housing Development and Management (HDM), Lund University.

Wamsler, C. and Brink, E. (2014), "Moving beyond short-term coping and adaptation", *Environment and Urbanization*, Vol. 26 No. 1, pp. 86-111.

World Bank (2007), "Dhaka: improving living conditions for the urban poor", Development Series, Paper 17, The World Bank, Washington, DC.

Further reading

Eriksen, S.H., Brown, K. and Kelly, P.M. (2005), "The dynamics of vulnerability: Locating coping strategies in Kenya and Tanzania", *The Geographical Journal*, Vol. 171 No. 4, pp. 287-305.

Moser, C. and Satterthwaite, D. (2008), "Towards pro-poor adaptation to climate change in the urban centres of low- and middle- income countries", Climate Change and Cities Discussion Paper [3], International Institute for Environment and Development (IIED), London.

Roy, M., Guy, S., Hulme, D. and Jahan, F. (2011), "Poverty and climate change in urban Bangladesh (ClimUrb): an analytical framework", BWPI Working Paper [148], Brooks World Poverty Institute, Manchester.

Tanner, T. and Mitchell, T. (2008), "Entrenchment or enhancement: Could climate change adaptation help to reduce poverty?", *IDS Bulletin*, Vol. 39 No. 4, pp. 6-13.

Corresponding author

Md. Ashiq Ur Rahman can be contacted at: tuhin_urp@yahoo.com

6

Migration as adaptation strategy to cope with climate change: a study of farmers' migration in rural India

Chandan Kumar Jha, Vijaya Gupta, Utpal Chattopadhyay and
Binilkumar Amarayil Sreeraman
National Institute of Industrial Engineering, Mumbai, India

Abstract

Purpose – This study aims to evaluate the link between climate/weather change and farmer migration in Bihar, India. The influence of cognitive conditions and climate-related stress on farmer migration decisions and the socioeconomic characteristics of migrating and non-migrating farm households are analysed. The focus is the role of migration in access to climate and agricultural extension services and the contribution of migration to enhanced farmer coping capacity.

Design/methodology/approach – A primary survey was conducted of farm households in seven districts of Bihar, India. Farmer perceptions of climate change were analysed using the mental map technique. The role of socioeconomic characteristics in farm household migration was evaluated using binary logistic regression, and the influence of migration on access to climate and agricultural extension services and the adaptive capacity of migrating households was investigated using descriptive statistics.

Findings – Climate-induced livelihood risk factors are one of the major drivers of farmer's migration. The farmers' perception on climate change influences migration along with the socioeconomic characteristics. There is a significant difference between migrating and non-migrating farm households in the utilization of instructions, knowledge and technology based climate and agriculture extension services. Benefits from receipt of remittance, knowledge and social networks from the host region enhances migrating households' adaptive capacity.

Originality/value – This study provides micro-evidence of the contribution of migration to farmer adaptive capacity and access to climate and agricultural extension services, which will benefit analyses of climate-induced migration in other developing countries with higher agricultural dependence. In addition, valuable insights are delivered on policy requirements to reduce farmer vulnerability to climate change.

Keywords Adaptation strategies, Climate change, Migration, Agriculture, Extension services, Socioeconomic conditions

1. Introduction

Human displacement in response to environmental shocks is not a new phenomenon. Hippocrates and Aristotle believed that the characteristics of the natural environment determined the human habitability of a region and, in turn, the characteristics of its inhabitants (Livingstone, 2000). Large disparities in income and living standards are major reasons for increased migration from rural to urban areas in developing countries such as India. Economists including Fei and Ranis (1964), Harris and Todaro (1970), Stark (1984), Lucas (1997) have provided theoretical support for this hypothesis of internal migration. Fei and Ranis (1964) broadly explained that rapid internal migration is a desirable process by which surplus labour at $MP_L = 0[1]$ is withdrawn from traditional agricultural occupations to provide cheap manpower to the modern industrial state, where $MP_L > 0[2]$. A country will remain in a middle income trap if it fails to make the transition from labour-intensive to capital-intensive production (Kohli *et al.*, 2011).

Agriculture has predominantly become a climate-sensitive sector in which surplus labour ($MP_L = 0$) has increased and emerged as a major contributing factor to internal migration to non-climate-sensitive sectors (where $MP_L > 0$). The Harris and Todaro (1970) model of economic development explains some drivers of rural urban migration. The basic postulate of this model is that migration is based on the expected income differential between rural and urban areas rather than the wage differential alone. Lucas (1997) subsequently supported this model but added that the understanding of the factors determining the urban component other than the wage differential remains poor. De Haas (2010) also noted that:

> [...] whether migration occurs crucially depends on the skills and knowledge of migrants and conditions in the specific economic sectors where they are likely to find employment both at the origin and destination.

New Economics of Labour Migration (NELM) models conceptually differ from other models by incorporating rural risk as a factor determining migration decisions. According to this approach, the migration decision is based on important human groups such as families and households rather than the individual alone (Mincer, 1978; Stark and Levhari, 1982; Stark and Bloom, 1985; Stark and Lucas, 1988). This model also views migration as a family strategy aimed at maximizing expected earnings and reducing the risk of consumption failure by diversifying income sources across sectors or agro-zones.

Human migration is not only a response to poverty and social deprivation but also an adaptive response to changing climate (Scheffran *et al.*, 2012). Extreme weather events, sea level rise and environmental degradation are major consequences of climate change. These changes are major causes of short- or long-term migration due to loss of climate-sensitive occupations and low adaptive capacity in South Asia, including India (Bhatta and Aggarwal, 2016; Kumar and Viswanathan, 2012; Deshingkar and Akter, 2009, Deshingkar and Start, 2003). Most developing countries face resource constraints in meeting the demands of high populations and thus have low adaptive capacity (Ahmed *et al.*, 2012). The adaptive capacity of a household is based on several factors, including financial resources, access to information, social resources, human capital and infrastructure (Barnett and Webber 2009). Households may reduce expenditures on non-essential goods, use formal and informal credit or draw on public assistance (Gray and Mueller, 2012). Alternatively, or in conjunction, families may send a member elsewhere to access alternative income sources for remittance to the origin household (McLeman and Smit, 2006). In the absence of group-based mitigation activities and spatial coping strategies, migration can be an important adaptation strategy for households to adapt to climate shocks. Due to uncertainty related to climate-sensitive occupations, households engage in livelihood diversification, a major coping strategy for various economic and environmental challenges (Bhatta *et al.*, 2015b). Remittances are the most obvious factor enhancing the capacity of households to adapt to stress (Banerjee *et al.*, 2011). The increase in income from diversified livelihood sources allows farmers to enhance their adaptive capacity to cope with climate risk (Cunguara *et al.*, 2011; Tripathi, 2017, Patnaik and Das, 2017).

Adaptation is a major policy option to avoid the dire consequences of climate change, and migration is one potential adaptation strategy (Nordas and Gleditsch, 2007; Laczko and Aghazarm, 2009; Tacoli, 2009; Scheffran *et al.*, 2012). As the most climate-centric economic activity, agriculture faces the greatest risk and uncertainties associated with loss of livelihood. Stojanov *et al.* (2016a, 2016b) found that at the national level, individuals first recognize change in climatic aberrations and choose to migrate as a coping strategy. Farmers are the main performers and actors at the farm level, and therefore farmer perceptions of climate change and the socioeconomic characteristics of farm households are the main factors driving farmer migration. Although perception of climate change is a prerequisite for the individual adaptation response, socioeconomic conditions determine the vulnerability of farm households to climate-induced economic tragedy and social deprivation. Farmers opt to migrate in search of alternative livelihoods under the risk of crop failure or low crop yield due to the impact of unpredictable climate consequences. In addition to generating a certain expected income, migration provides opportunities to acquire knowledge on new techniques and farming practices to enhance agricultural income. Moreover, remittances from the host region reduce credit constraints on the adoption of adaptation strategies (Patnaik and Narayanan, 2015), thus scaling up the adaptive capacity and resilience of the home region.

Given this theoretical background, there is a need to study the migration-climate nexus in a more context-specific manner by integrating multiple socioeconomic, political, cultural and developmental factors. Furthermore, there is a need to reveal how household characteristics influence the migration decision to develop more informed regional policies (Upadhyay *et al.*, 2015). This study develops an understanding of the role of migration as an adaptation strategy to cope with climate change and attempts to reframe the debate as "migration is not always the result of an individual failure to adapt effectively". The study assesses whether farmers in the study area perceive changes in climate/weather and accordingly identifies different cognitive conditions inducing the decisions of farmer to migrate for the Kharif or Rabi season. Furthermore, the socioeconomic conditions of migrating and non-migrating farm households are evaluated. The focus of this study is to evaluate the role of migration in access to climate and agricultural extension services and further assess the contribution of migration in enhancing farmer coping capacity.

The remainder of the paper is organized as follows. Section 2 briefly reviews the literature on different causes of migration, including climate-induced migration. Section 3 discusses the study area, and Section 4 describes the methodology and survey design. Section 5 presents the results and discussion of farmers' perceptions of climate change and their migration decisions, the role of socioeconomic characteristics in migration as an adaptation strategy, the role of migration in the choice of extension services, migration as an adaptation strategy and the adaptation strategies adopted by migrating and non-migrating farmers. Finally, Section 6 discusses the conclusions and policy implications of the study.

2. Literature review

Climate change is a stress factor underlying migration in climate-vulnerable regions (Warner and Afifi (2014); Adger *et al.*, 2002, 2015; Curtis and Schneider, 2011; Gray and

Bilsborrow, 2013; Henry *et al.*, 2004; Hunter *et al.*, 2013; Mueller *et al.*, 2014; Stal, 2011). Migration provides individuals with an opportunity to make non-marginal adjustments in adapting to climate change (Klaiber, 2014), and the potential impacts of climate change on migration have long been a subject of intense interest in the policy and academic arenas (Warner, 2010; Renaud *et al.*, 2007; Stern, 2007; Conisbee and Simms, 2003).

A baseline assumption prevalent in the literature is that individuals migrate to other locations because they failed to adapt effectively (Warner, 2010; Renaud *et al.*, 2007; Conisbee and Simms, 2003; Adamo, 2010). Environmental factors indirectly affect individual migration decisions by impacting economic activity, e.g. loss of agricultural productivity and increased expenditure due to increasing food prices (Porter *et al.*, 2014). Credit sources may be insufficient for the adoption of modern technologies for climate change mitigation and adaptation (Lybbert and Sumner, 2012). In most South Asian countries, the probability of migrating under adverse conditions is higher among households with poor socioeconomic profiles (Etzold *et al.*, 2014; Bhatta *et al.*, 2015a; Bhatta *et al.*, 2015b). Migration-related decisions are based on vulnerability to economic, political, social and demographic factors (Stojanov, 2004; Stojanov *et al.*, 2016a, 2016b; Yi Sun *et al.*, 2017). Families may relocate individual members to earn income to sustain the expenditures of the family, acquire knowledge and increase capabilities to confront future shocks and stresses (de Haan *et al.*, 2002).

2.1 Reasons for migration

NELM (Taylor, 1999) describes migration as a potential source of capital transfer to help traditional communities move towards modern knowledge and education. For instance, agricultural extension is an important engine of knowledge innovation and development in developing countries (Rivera and Sulaiman, 2009). Extension services are comprehensive institutional arrangements that help farmers organize themselves and links farmers to markets (Swanson, 2006). Transfer of resources to the home region, including knowledge, remittance and migrant return, can contribute to increased awareness of the value of technical and institution sources of climate information and extension services. Remittance income directly affects the resource base, economic well-being and resilience of the home community (Adger *et al.*, 2002). Individual migrants scale up community capabilities to acquire sustainable livelihoods and development. In developing countries, remittance income is increasing substantially due to high rates of migration. These links between host and home communities are required to strengthen the resilience of the home community (Conway and Cohen, 1998), increase its economic well-being (Adger *et al.*, 2002) and provide access to resources that enrich its human, social and cultural capital (Woodruff and Zenteno, 2007, Massey *et al.*, 1993; Taylor, 1999). Social networks in the host region have also identified as a major factor for migration by providing support in terms of accommodation and employment opportunities (Ravuvu, 1992).

2.2 Migration due to climate change

A broad spectrum of the literature supports links between climate change and prehistoric human settlement and migration (Huntley, 1999; Tyson *et al.*, 2002). In response to the loss of climate- sensitive occupations and the non-availability of climatic-insensitive occupations, communities diversify their livelihoods by intensifying agricultural and non-farm activities (McDowell and Haan, 1997). Tschakert and Tutu (2010) further emphasized the importance of migration in coping with climate change; in particular, in South Asia, a large number of landless and marginal farmers have migrated to cope with climate variability (Bhatta and Agarwal, 2016). The impact of climate change on human migration dynamics is particularly notable in rural areas due to a lack of adaptive capacity (Boyd and Ibarrarán, 2009; Kates, 2000). Perception of climate change is an important precondition for individual adaptation responses, including farm-level adaptation (Bryan et al., 2009; Hassan and Nhemachena, 2008; Gbetibouo, 2009). At the micro-level, climate-driven migration decisions of farm households initially depend on their perception of climate change/variability, the associated risks and uncertainties of crop failure and loss of livelihood and the need for alternate income sources. A farmer's decision to migrate follows a cognitive process involving recollection of past incidents of climate extremes and current understanding of the climate based on intuition and perceptions of risks associated with climate change (Grothmann and Patt, 2005; Adger et al., 2009; Marx et al., 2007). However, given the significant differences between personal experience and external sources of information due to climate uncertainty, farmers may place greater weight on recent climatic events (Hansen et al., 2004). Wider gaps between climate variations anticipated by farmers and actual climate forecasts may also lead to indecisiveness (Roncoli et al., 2002; Below et al., 2010). Changes in agricultural systems do not involve linear updating of a farmer's decision making, and therefore the manner in which farmers revise their anticipations of climate variation crucially determine adaptation decisions (Bryan et al., 2009; Gbetibouo, 2009).

Despite the plethora of literature on climate change, adaptation and migration, gaps remain in identifying the prevailing heterogeneous influences in the home region that drive migration and how migration further enhances the migrating member's ability to acquire knowledge on farm management and extension benefits. Studies of climate-induced migration patterns among different heterogeneous groups are very limited (Bhatta et al., 2015b). Therefore, it is pertinent to understand migration as the major source of household resilience under the scenario of climate change.

3. Study area

This study was conducted in three sub agro-climatic zones of the Middle Ganga Plain in the state of Bihar, India. The river Ganga divides the state into North Bihar (area of 53,300 km²) and South Bihar (area of 40,900 km²). North Bihar (Zones I and II) is highly flood prone, whereas South Bihar (Zone III) is drought prone (Agricultural profile of the state, BAMETI, 2017). Annual precipitation in the state varies between 990 and 1,200 mm, with major precipitation during the months of July to September. The state experiences the summer season between the months of April and June and the winter season from December to February. The hottest month is May, with a maximum temperature of 45°C, and the coldest month is January, when the temperature falls below 10°C (IMD, 2009). These sub agro-climatic zones have experienced the brunt of changing temporal and spatial patterns of rainfall and temperature. The state has two major cropping seasons, Kharif and Rabi[3]. The population pressure is quite high in the state, with a population density of 1,102/km², in contrast to the national average of 382/km² (Census of India, 2011). Consistent with the high population density, resource constraints are enormous in Bihar: 42.6 per cent of the population is below the poverty line, in contrast to the national average of 26.10 per cent (Department of Agriculture, Govt. of Bihar). Approximately 89 per cent of the population resides in rural areas where agriculture and animal husbandry are the main sources of livelihood.

Internal migration is a major issue in the Indian economy due to the history of unequal growth among states. The availability of migration data in India is limited to two sources: census data collected by the Registrar General of India and survey data (employment-unemployment surveys or special migration surveys) collected by the National Sample Survey Organization (NSSO). Bihar is a major contributor to internal migration due to high rates of temporary and seasonal migration (Keshri and Bhagat, 2013). The major reason for migration is the unavailability of non-farm employment opportunities and other attributes of higher living standards. According to the 2011 census, Bihar has the second highest rural population, 11.1 per cent, after Uttar Pradesh, which has a rural population of 18.6 per cent. Bihar also ranks at the top of migrant-sending states in India.

The Economic Survey of India (2016-17), which uses the new Cohort-based Migration Metric (CMM)[4], reports that annual average inter-state labour mobility averaged 5-6 million people between 2001 and 2011 and accounted for 60 million inter-state migrants and 80 million inter-district migrants. The first-ever estimates of railway passenger traffic data for the period 2011-2016 reveal work-related migration of approximately nine million people, nearly double the figures in the 2011 Census (Economic Survey, GOI 2017). Based on the relationship between CMM scores and per capita the survey reveals that the rate of net out-migration in less developed state such as Bihar and Uttar Pradesh is high.

Different data sources have classified Bihar as one of the highest contributors to out-migration in India. The combination of natural, economic, and social circumstances forces family members to migrate. The NSSO 64th round also highlighted Bihar as a major migrant-sending state and identified work/employment and education as the major reasons for male migration and marriage as the major reason for female migration. Domestic remittances from the host region to meet household consumption expenditures are a major driver of internal migration in Bihar. Tumbe (2011) found that the dependence on domestic remittances is much higher in Bihar than the average for India and has grown since the 1990s.

4. Methodology

4.1 Sample selection

This study was conducted in Bihar state of India during the first half of 2015. The state of Bihar is divided into three sub-agro-climatic zones of middle Ganga plain on the basis of different hydrological, climatic, soils and agricultural attributes. The 7 districts were randomly selected out of 38 districts from the three sub-agro-climatic zones. The 72 villages from these 7 districts were selected randomly which were closer to the district headquarter and farm household have easier access to inputs, institutional farm aid and management and vice versa. The district headquarters have villages within the periphery of 10-70 kilometres (subject to size of the districts) with their institutional support and market access. From these 72 villages, 735 farm households were selected randomly. Finally, the data from 700 farm households was found complete in all aspects for the purpose of statistical analysis. This study included all land size groups, such as marginal, medium, small and large. The preliminary information on the farm household was collected from the office of Head of the village.

4.2 Questionnaire design

The self-administered structured questionnaire was discussed with the farm household to ensure consistency and collect data. If a household declined to respond, the next farm household in the same land size category was approached. The selected households were engaged in farming either on their own farms (i.e. land ownership) or others' farms (i.e. rented land). Therefore, all households qualified for consideration as farmers or farm households because they were the main on-farm actors and decision-makers. The questionnaire was designed to capture the farmers' local perceptions and observations of climate change and experiences with climate variability and extreme events over the past two decades. In addition, a mental map technique was adopted in which the famers were asked about their local perceptions of changes in climate based on their past and current experiences and observations of changes/variations in climate variables. Ethnographic studies have confirmed that individuals can correctly identify climate change over a decade or longer based on personal experience (Marin, 2010; West *et al.*, 2003; West *et al.*, 2008). For example, a recent study in Burkina Faso found that farmers successfully recognized a decrease in rainfall that occurred over a 30-year period (West *et al.*, 2008). Furthermore, the farmers were enquired about their socioeconomic conditions, their choices and access to information about climate and new agricultural techniques (climate and agricultural extension services) and, accordingly, their choices of adaptation strategies among 11 different adaptation alternatives[5] for the Kharif and Rabi seasons separately[6]. The adaptation options attempted to capture capital-intensive, labour-intensive and knowledge-intensive techniques to facilitate the assessment of whether migration contributes to the adoption of adaptation strategies.

4.3 Data analysis
A binary logistic regression model was used to assess the role of socioeconomic characteristics. The study was based mainly on the assumption that the migration decisions of farmers are family strategies and therefore either the farmer or at least one member of the household will migrate to nearby zones or more distant cities. Furthermore, the study principally used behavioural attributes of migrating and non-migrating farmers and did not include psychological factors other than perceptions of climate stress as drivers of farmer migration. Descriptive statistics were used to reveal differences in adaptation behaviour, choice of agriculture and climate extension services and pattern of adoption of different adaptation strategies between migrating and non-migrating farmers or farm households.

5. Results and discussion
5.1 Perceptions of climate change and migration
A farmer's experiences and perceptions of climate variability and climate extremes determine the nature of migration, i.e. temporary or permanent. For instance, farmers who are more vulnerable to climate extremes and experience continuous changes in climate patterns are more likely to permanently migrate due to continuous loss of assets (e.g. land), wealth and livelihood, whereas farmers affected by seasonal climate variations and provisional changes in climate may decide to make temporary income arrangements to fulfil livelihood requirements. Temperature and precipitation were considered as climate variables in this study. The farmers' perceptions of climate change were evaluated based on their responses (i.e. yes or no) to the question "Have you observed any long-term changes in mean temperature and rainfall levels over the last 20 years?". Of the surveyed farmers, approximately 91 and 86 per cent reported observing changes in temperature and rainfall patterns, respectively, over the past 20 years.

The decision to migrate was analysed for two cropping seasons, Kharif and Rabi, because climate conditions for ideal crop productivity and thus farm expenditures and management practices differ by farming season. Based on the farmers' responses on the perception of changes in temperature and precipitation and their choices to migrate due to crop sensitivity, four cognitive conditions of the farmers' decision were considered:

  (1)   perceive changes in climate and migrate;

  (2)   perceive changes in climate but do not migrate;

  (3)   do not perceive changes in climate and therefore do not migrate; and

  (4)   do not perceive changes in climate yet choose to migrate.

Table I presents the farmers' decisions to migrate as driven by their perceptions of climate change/variability. Of the 700 farm households surveyed, based on the sensitivity of planted crops to temperature changes, approximately 62 and 38 per cent opted to migrate and not migrate, respectively, in Kharif; in Rabi, approximately 64 per cent opted to migrate and 36 per cent chose not to migrate. Furthermore, based on the sensitivity of planted crops to changes in rainfall conditions, 60 and 40 per cent opted to migrate or not migrate, respectively, in Kharif; in Rabi, approximately 67 per cent opted to migrate, and 33 per cent chose not to migrate. In both seasons, the percentage of respondents opting to migrate due to perceived changes was much higher than the percentage choosing to migrate despite not perceiving climate change. However, a high percentage (30-35 per cent) of respondent perceived climate change but did not migrate. The characteristics of the farm households can explain these differences in choice.

The majority of the surveyed farmers considered migration a profitable coping strategy. Perceiving climate change and migration was the dominant cognitive condition for both Kharif and Rabi in the study area. This condition is mainly driven by the rural risk factors associated with climate changes/variations in the home region. The decision to migrate is part of a household coping strategy rather than an individual decision. Farmers under the second cognitive condition, i.e. migrating despite not perceiving climate change, mainly aim to diversify and maximize their income sources. A few farmers chose not to migrate despite perceiving climate changes (the third cognitive condition), mainly because of uncertainties in the host region (e.g. guarantee of employment, costs associated with migration) and unfavourable socioeconomic conditions in the home region (e.g. household size, education, land size). The fourth cognitive condition implies that farmers are not aware of any changes in climate conditions and are either satisfied with their economic conditions in the home region, do not find migration a profitable option or lack the skill and knowledge to migrate.

5.2 Role of socioeconomic characteristics in migration as an adaptation
An understanding of the socioeconomic characteristics of farmers is important to recognize the main factors driving migration. The migration decisions of farmers are family strategies that are mainly undertaken to overcome the risk of loss of income and of social deprivation and ensure consumption smoothening. To assess the role of socioeconomic conditions in farmer migration as an adaptation, a binary logistic regression model was used with the migration decision of farmers ("1" for migration and "0" for non-migration) as a single categorical predictor. Socioeconomic variables such as age and education level of the farm household head; proportion of males in the household to total household size; proportions of other family members, i.e. proportion of females and children younger than 10 years of age;

Table I. Perceptions of climate change/variability as drivers of farmers' decisions to migrate

Conditions	Kharif season		Rabi season	
	Temperature	Precipitation	Temperature	Precipitation
Migrated total	62	60	64	67
Not migrated total	38	40	36	33
Perceive and migrated	57	52	59	58
Did not perceive but migrated	5	8	5	9
Perceive and not migrated	34	34	33	29
Did not perceive and not migrated	3	6	3	4

Notes: The numbers are in percentages and they may not add up to 100 due to rounding off
Source: Own calculation

size of land; land ownership pattern (own farm land or rent); and livestock were considered as independent variables.

As shown in Table II, age and education of the farm household head, male proportion, proportion of other family members, farm size and land ownership were significant factors in Rabi. The results for Kharif were similar, except farm size. Age of the farm household head was positively significant for both seasons. Farmer age is a measure of farm experience, which determines a farmer's perception, willingness to adapt and adaptation decisions. Older farmers feel a greater sense of responsibility for ensuring sustained livelihood sources for their families and therefore either migrate themselves or encourage other household members to migrate, not only to reduce dependency but also to ensure remittances to the home region. As noted earlier, the dependence on domestic remittances has increased in Bihar since the 1990s (Tumbe, 2011). Education level positively affected the migration of farmers or household heads by increasing their capability to acquire and synthesize information on climate conditions and respond; farmers with more education are also more aware about available opportunities in the host region. Furthermore, the intra-household spill-over effect of education encourages other household members to move out of the home region as well. In the study area, a higher proportion of males in the household positively influenced the intent to migrate, as more male members ensure more earning hands and more efficient exploitation of social networks in the host region. The proportion of women and children younger than 10 years of age signifies dependency; therefore, a higher proportion positively influenced migration decisions because of the need for income to meet higher consumption requirements. Land ownership pattern was negatively significant for both seasons, implying that farmers who do not possess their own land are more likely to migrate as a result of lower net income (net agricultural income after paying land rent) that is insufficient to meet consumption requirements and other household expenditures in the home region. Farm size was negatively significant in Rabi. The larger the land size, the greater the requirement for family labour for farm management; consequently, farm households may not intend to migrate. However, farm size was not significant for Kharif.

5.3 Role of migration in extension services

The present scenario of climate change and its related risks demands regular access to new knowledge through adequate and well-timed extension services. Agricultural extension services and credit facility are the two institutional arrangements that most enable adaptation. Agricultural extension facilitates the delivery of information on seasonal climate variations and new technologies to help farmers perceive climate changes quickly and modify their agronomic practices accordingly. Migration is a major source of awareness via

Table II. Results of binary logistic regression model

Variables	Migration in Rabi season	Migration in Kharif season
Age (Head of the HH)	0.0493628 (4.00)**	0.079568 (5.73)**
Education (Head of the HH)	0.1590579 (5.09)**	0.1506099 (4.45)**
Farm size	−0.678732 (−2.07)*	0.0442645 (1.39)
Ownership patterns	−0.678732 (−2.62)*	−0.7613176 (−2.72)*
Male proportion to HH size	5.847889 (5.20)**	7.232698 (5.85)**
Other proportion to HH size	1.202599 (8.54)**	1.759723 (10.30)**
Livestock	−0.016438 (−0.20)	0.009304 (0.11)

Notes: * $p < 0.05$; ** $p < 0.01$
Source: Own calculation

income and knowledge transfer from the host region to the home region for efficient utilization of extension services. Through social networks, returning migrants help local communities advance the use of technology in agricultural activity and increase awareness of institutional arrangements that can enhance agricultural returns.

In the study area, farmers obtain climate-related information such as expected seasonal rainfall and temperature levels, timing of monsoon onset and predicted climate extremes such as flood and drought from climate extension sources. Agricultural extension services provide information on farm mechanization, new crop varieties, drought-tolerant crop varieties, availability of quality seeds, plant protection, soil health and market information. Climate extension sources include information from the meteorological department, local government sources, mobile (e.g. internet, telephone help lines), radio and television. Agricultural extension sources are identified as information from field officers, television, mobile, radio and television. Differences in the adoption of agriculture extension sources between migrating and non-migrating farm households during the different cropping seasons were assessed (Figures 1 and 2).

Migrating farm households made greater use of both climate and agricultural extension services than non-migrating households in both seasons (Figures 1 and 2). Among migrating farmers, 70-77 per cent still used radio, which is the oldest source of

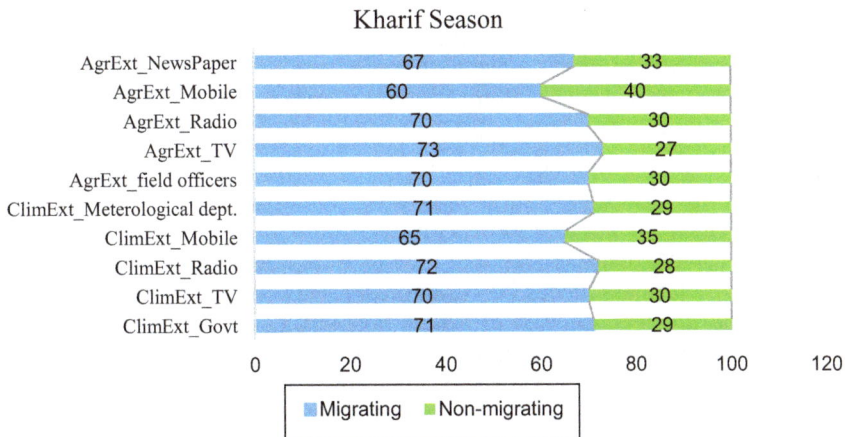

Figure 1. Choice of extension services by migrating and non-migrating households in Kharif

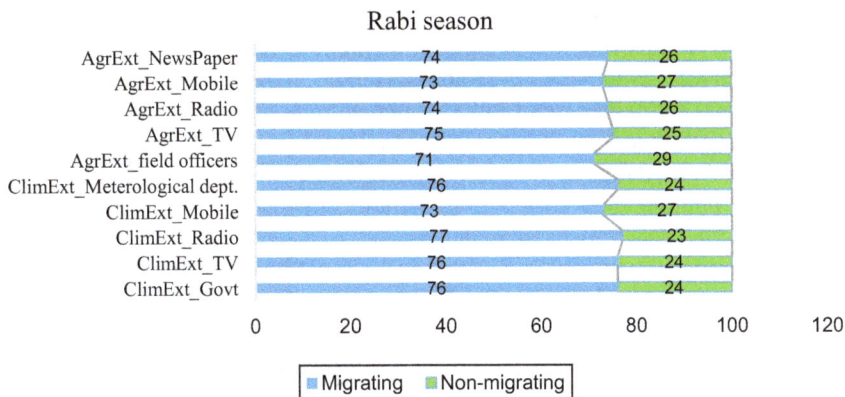

Figure 2. Choice of extension services by migrating and non-migrating farm households in Rabi

information on climate and agriculture; approximately 60-73 per cent used mobile; and 67-74 per cent used newspapers. Furthermore, approximately 71-76 per cent of migrating farmers found climate information provided by the meteorological department and field officers useful. Migrating households are better informed about the use of sources such as mobile and television in agricultural management based on experiences gained from their social networks and experiences in the host region (Scheffran *et al.*, 2012). The utilization of other extension sources, such as newspapers, depends upon the education level of the farmers. Because mobile is a knowledge-based technology, migrating farmers are more efficient in utilizing these resources for agricultural management. Also, remittances from the host region improve economic well-being, which enables the purchase and utilization of televisions, mobile phones, newspapers, computers and radios in the home region. The non-migrating farmer's need for information can also be met by sharing information, knowledge, technology and skills in the migrating communities, thus improving "social learning" and the adaptive capacity of the home region.

5.4 Migration as an adaptation strategy or source of adaptation strategy
This section examines how migration enhances a farmer's adaptive capacity and enables the adoption of other adaptation strategies. In all, 11 different adaptation strategies (including migration) were identified as the most suitable farm management practice, farm technology and farm financial management alternatives: planting different crop varieties, changing land under cultivation, increasing/decreasing irrigation, adopting soil conservation techniques, adopting water conservation techniques, crop insurance, migration, switching from farm to non-farm activities, leasing land, planting horticulture and planting vegetables. Farmers expressed their choices of adaptation strategies in accordance with their perceptions of changes/variations in temperature and rainfall and their own farming practices separately for Kharif and Rabi. Although these strategies may be profit-driven instead of climate-driven, their choices of strategies were assumed to be driven by climatic factors as reported by farmers themselves (Maddison, 2006; Nhemachena and Hassan, 2007).

To examine the farmers' choices of adaptation strategies, the adaptation level was developed based on the number of adaptation strategies adopted (Bhatta *et al.*, 2016). The number of adaptation strategies reported by the farmers ranged from a minimum of 0 (no adaptation considered) to 11 (maximum number of adaptation strategies used by any farmer). The average number of adaptation strategies adopted by farmers in Kharif for perceived changes in temperature and rainfall was approximately four for both. In Rabi, the average was approximately five for both perceived changes in temperature and rainfall. Based on these averages, four adaptation scores were identified: no adaptation (0), low adaptation level (1-3 adaptation strategies), medium adaptation level (4-6 adaptation strategies) and high adaptation level (7-11 adaptation strategies). The farmers reported their choices of adaptation strategies separately for perceptions of temperature and rainfall and for Rabi and Kharif. Therefore, the choices of adaptation strategies were not mutually exclusive as farmers could decide to make several changes at the farm level simultaneously (e.g. farmers may change crop varieties, irrigation and insurance simultaneously due to flood risks or shortage of rainfall in Kharif).

Adaptive capacity is the property of a system to adjust its characteristics or behaviour to expand its coping range under existing climate variability or future climate conditions (Brooks and Adger, 2005). Accordingly, this study assumes that farmers with a higher adaptation level (coping range) have higher adaptive capacity. Adaptive capacity also demonstrates individual capacity to moderate the impacts of climate change at the farm level as finely determined by the skill, education and personal ability of the farmer (Tarleton and Ramsey, 2008). Table III presents the results for the importance of migration in enhancing a farmer's adaptation level (adaptive capacity) by cropping season in the study area. Among the 700 farmers surveyed, 138 (20 per cent) and 136 (19 per cent) did not adapt to changes in temperature and rainfall, respectively, in Kharif; in Rabi, 74 (11 per cent) farmers did not adapt to changes in both.

Among the surveyed households, 60-62 per cent and 64-67 per cent adapted migration as an adaptation strategy in Kharif and Rabi, respectively. More farm households chose to migrate in Rabi than in Kharif, which is more labour-intensive and requires numerous farming activities. Due to the lack of adequate work for all members of the household in Rabi, some chose to migrate. According to Viswanathan and Kumar (2015), out-migration from one state to another in India is more sensitive to changes in rice yield (a Kharif crop) than wheat yield (a Rabi crop). This discrepancy with the results of the present study is mainly attributable to differences in scale and the type of data used to assess migration. Viswanathan and Kumar (2015) focused on state-level migration in India using actual weather, crop yield and per-capita net state domestic product. By contrast, the present study is a micro-level evaluation of migration as an adaptation strategy and mainly uses uncertain and unpredictable behavioural attributes of farmers. In Kharif, for changes in temperature and rainfall, 204 (29 per cent) respondents had a low level of adaptation, 279 (40 per cent) farmers had a medium adaptation level, and only 81 (12 per cent) had a high level of adaptation. Similarly, for Rabi, for changes in temperature, 171 (24 per cent) respondents had a low level of adaptation, 306 (44 per cent) had a medium adaptation level, and only 149 (21 per cent) had a high level of adaptation; for rainfall changes, 174 (25 per cent) had a low level of adaptation level, 305 (44 per cent) had a medium adaptation level, and only 147 (21 per cent) had a high level of adaptation. Thus, in the study area, a medium level of adaptation was most prevalent, and farmers mostly adopted four to six different adaptation strategies in a season in response to changes in temperature or rainfall.

In Kharif, among the 700 surveyed farmers, 437 (62 per cent) and 418 (60 per cent) opted for migration due to changes in temperature and rainfall, respectively. In Rabi, 450 (64 per cent) farmers and 471 (67 per cent) farmers chose to migrate due to changes in temperature and precipitation, respectively. The importance of migration in enhancing a farmer's adaptation level (adaptive capacity) by cropping season in the study area was examined by evaluating the proportion of migrating farmers to total adapting households at each level of

Table III. Importance of migration in enhancing a farmer's adaptation level (adaptive capacity) by cropping season in the study area

Household (HH) adaptation (in terms of no. of adaptation strategies)		Adaptation level			
	No (0)	Low (1-3)	Medium (4-6)	High (7-11)	Total
Kharif season					
Total No. HH (Temp)	138 (20)	171 (24)	300 (43)	91 (13)	700 (100)
Migrating HH (Temp)	0	116 (57)	248 (89)	73 (90)	437 (62)
Total No. of HH (Rain)	136 (19)	204 (29)	279 (40)	81 (12)	700 (100)
Migrating HH (Rain)	0	116 (57)	238 (85)	64 (79)	418 (60)
Rabi season					
Total No. HH (Temp)	74 (11)	171 (24)	306 (44)	149 (21)	700 (100)
Migrating HH (Temp)	0	78 (46)	240 (78)	132 (89)	450 (64)
Total No. of HH (Rain)	74 (11)	174 (25)	305 (44)	147 (21)	700 (100)
Migrating HH (Rain)	0	88 (51)	253 (83)	130 (88)	471 (67)

Source: Author's own calculation, Figures in parenthesis are shows percentages of total

adaptation. In Kharif, approximately 79-90 per cent of farm households with a high level of adaptation and approximately 85-89 per cent of farm households with a medium adaptation level were migrating households. Similarly, in Rabi, approximately 89 per cent of farmers with a high adaptation level and 78-83 per cent of farmers with a medium adaptation level were migrating households. These results indicate that migrating households have a higher adaptation capacity as they are able to adopt more adaptation strategies due to support from host regions in the form of remittances, knowledge, resources and networks.

5.5 Adaptation strategies adopted by migrating farm households by season in the study area
To assess the importance of migration in the adoption of other available adaptation strategies in the study area, the adaptation strategies adopted by migrating and non-migrating farm households were analysed separately for observed changes in temperature and rainfall and for Kharif and Rabi (Figure 3).

For both Kharif and Rabi, the responses for changes in temperature and rainfall were similar: migrating households reported choosing more adaptation strategies than non-

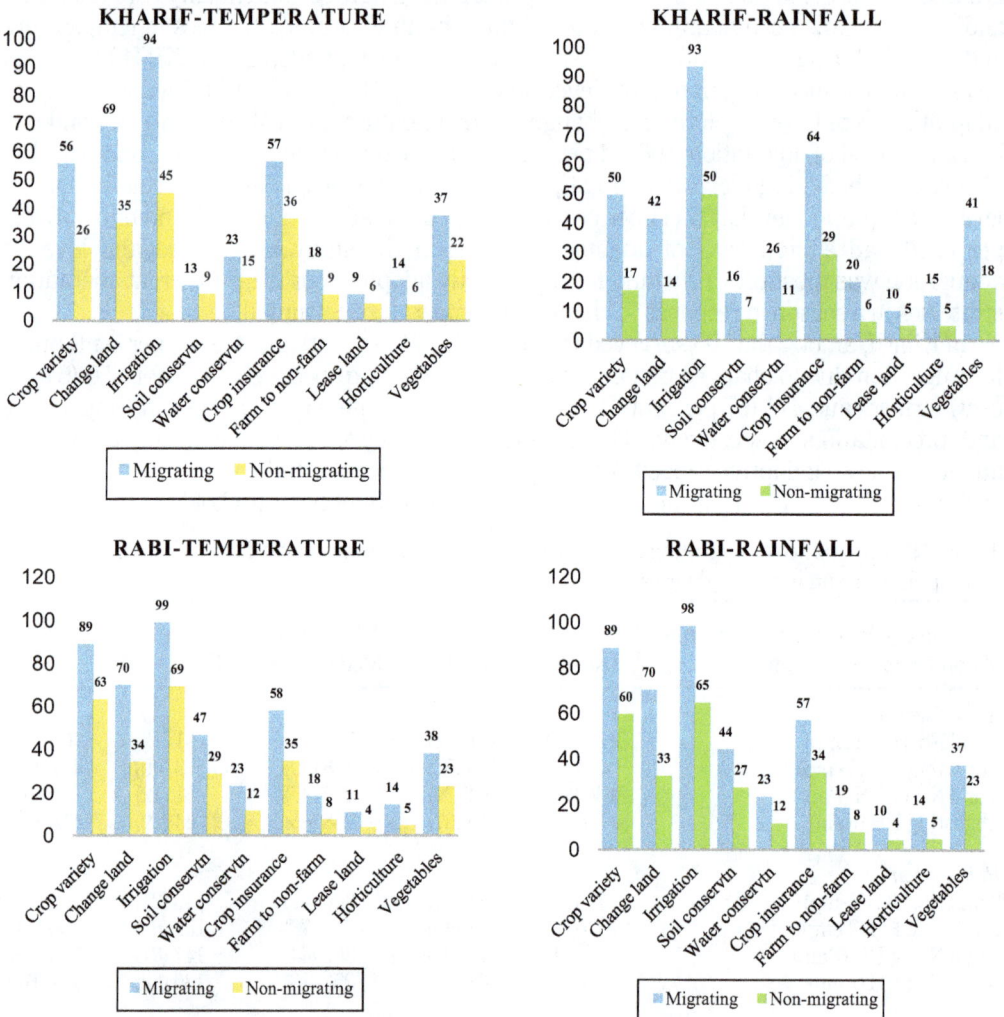

Figure 3. Adaptation strategies of migrant and non-migrant households in Kharif and Rabi

migrating households. In Kharif, for temperature and precipitation, 94 per cent of migrating farm households chose irrigation, 42-69 per cent opted to change land under cultivation, 50-56 per cent changed crop variety, 57-64 per cent opted for crop insurance, 23-26 per cent chose water conservation, 13-16 per cent chose soil conservation and 37-41 per cent planted vegetables. Similarly, in Rabi, 99 per cent chose irrigation, 70 per cent opted to change land under cultivation, 89 per cent changed crop variety, 58 per cent opted for crop insurance, 23 per cent chose water conservation, 44-47 per cent chose soil conservation and approximately 38 per cent planted vegetables. Compared with migrant farm households, the proportion of non-migrating farmers that chose these strategies was remarkably low.

For instance, in Kharif, only 45-50 per cent of non-migrating farmers chose irrigation, 29-36 per cent chose insurance and 7-9 per cent opted for soil conservation. Similar responses were recorded for Rabi; only 65-69 per cent chose irrigation, approximately 34 per cent chose to change land under cultivation, and 12 per cent opted for water conservation. The most common adaptation strategies chosen by migrating and non-migrating farm households for all four scenarios were changing crop variety, changing land under cultivation, irrigation, crop insurance, soil and water conservation and planting vegetables. These adaptation strategies are knowledge-, capital- and resource-intensive in nature, and therefore migrating farm households has a comparative advantage over non-migrating farm households. Migrating farm households benefitted from the remittances and knowledge they gained from the host region. Migrant farmers gained knowledge about crop varieties, new methods of irrigation and soil and water conservation techniques from the migrant network in the host region. Migrating farmers were more informed about new government initiatives for crop insurance schemes and new credit facilities and thus were able to utilize these facilities. In addition, remittances from the host region helped migrating farmers buy seeds for new crop varieties and adopt advanced farm mechanization techniques. The results indicate that remittances, knowledge and social networks from the host region directly enhance the economic conditions of migrating households and, to some extent, non-migrating households, further enhancing their ability to adapt in the home region. Thus, at the micro-level, farmer migration can be beneficial for improving adaptation, resilience and livelihood sustainability in the home region.

6. Conclusions

Climate-induced migration has typically been considered a livelihood or survival strategy under adverse and extreme climate conditions. Although farmers clearly often resort to migration either to diversify income sources or smooth consumption, migration has a much wider role to play as an adaptation strategy. The present study evaluated the influence of climate change perception on farmer migration. In addition, the contribution of migration in enhancing the adaptive capacity of farmers by scaling up their financial capabilities and knowledge of modern agricultural technology was analysed. Furthermore, the influence of migration on accessing climate and agricultural extension services was explored, and differences in adaptation strategies for the two different seasons of Kharif and Rabi were characterized to identify improved cultivation practices that are more tolerant towards climate change. This study finds that climate-induced livelihood risk is the major driver of farmer migration for both seasons. Migration is also related to economic incentives in the form of increased employment opportunities and to the value of the crop at home. For example, in Kharif, farm households choose to migrate less due to the cultivation of high-value crops and labour intensive such as rice that require more labour for farm management. By contrast, due to a lack of adequate work in Rabi, some family members may opt to migrate. With respect to socioeconomic conditions, the age of the head of household, proportion of male members in the household and the dependency ratio enhance the sense of responsibility towards sustaining household livelihood sources and therefore positively influence farmer migration. Educated farmers are more inclined to migrate as they are acquainted with benefits and opportunities available elsewhere. Furthermore, land size has a negative influence on farmer migration. Ownership is also an important factor in migration. A large farm size and ownership of the land increase agricultural income, which can aid consumption smoothening and other needs. By comparison, marginal and small-holding farmers and farmers operating on rented land have a higher sense of insecurity and therefore higher pressure to migrate.

Compared with non-migrating households, migrating farm households admit better utilization of instruction-, knowledge- and technology-based climate and agriculture extension services such as information from meteorological divisions, agricultural field officers, mobile, newspapers and television. Migrating farm households mostly have a higher adaptation level and are more capable of adopting knowledge-, capital- and resource-intensive adaptation strategies due to the receipt of remittances from migrating household members. Non-migrating households receive positive externalities due to spillover. Overall, the analysis indicates that migrating households have a comparative advantage over non-migrating households in terms of adaptive capacity. To further understand migration as an adaptation strategy, net revenue differences between migrating and non-migrating households should be explored.

In terms of policy perspectives, the focus should be on developing the nexus between risk reduction, development and capacity building in rural areas. The government should strive towards creating non-climate-sensitive livelihood options to diversify farm household income and enhance adaptive capacity. In addition, capacity building and training on knowledge-intensive adaptation strategies are needed in rural areas. Policies must be developed that nurture and strengthen social networks through community participation to enhance group-based adaptation approaches.

Notes

1. A marginal product of labour (MP_L) equal to zero indicates that the last added unit of labour has zero productivity and that the marginal product is not equal to a subsistence level of wages. Lewis defined this level as "disguised unemployment".

2. A marginal product of labour (MP_L) greater than zero indicates that the last added unit of labour has positive productivity and that the marginal product is at least greater than the subsistence level of wages.

3. The cropping season in Bihar is classified into two main seasons based on weather attributes: Kharif and Rabi. The Kharif cropping season is from July to October during the southwest monsoon, and the Rabi cropping season is from October to March (winter). The crops grown between March and June are summer crops.

4. CMM (t) = 100 × [Population in 20-29 age cohort in Census(t) – Population in 10-19 age cohort in Census(t-10) – Cohort Mortality]/[Population in 10-19 age cohort in Census(t-10), Cohort mortality 10 x Age-specific (10-19) mortality rate per year × Population in 10-19 age cohort in Census(t-10)]. Data Source: Population data from the 1991, 2001 and 2011 Census and age-specific mortality data (State level) for the 10-19 age group for the years 1996 and 2006 from Sample Registration System statistics.

5. Changing crop varieties (planting different crops, drought-resistant varieties, high-yield verities, water sensitive crops, short duration varieties); changing land under cultivation (land rotation or altering the area under cultivation); increasing irrigation (increase/decrease the intensity of

irrigation to overcome shortage or excess of rainfall. Using tube well, water pump, etc.); adopting soil conservation strategies (for maintaining soil fertility-like zero-tilling, etc.); adopting water conservation strategies (rain water harvesting, building tanks or water reservoirs); crop insurance (insure crops to overcome crop losses due to climatic disturbances); migration (migrating to urban area to diversify their livelihood options); switching from farm to non-farm activities (changing land from farm to non-farm activities mainly in non-climate sensitive activities); leasing land for other purposes (leasing land for other non-farm activities; planting horticulture crops along with major crops (planting fruits like- mango, litchi, banana, guava etc. and nuts, seeds, herbs, sprouts, mushrooms, flowers, seaweeds and non-food crops such as grass and ornamental trees and plants); planting vegetables (Planting vegetables like potato, brinjal/eggplant, cauliflower, cabbage, tomatoes, chili, etc.)

6. Considering the high rate of temporary and seasonal migration in Bihar (Keshri and Bhagat, 2013), migration was considered as an option in both Rabi and Kharif separately. In this sense, work, business, and education are broadly considered as reasons for migration. Marriage and place of birth are not considered as reasons for migration because these reasons are based on social considerations (Viswanathan and Kumar, 2015).

References

Adamo, S.B. (2010), "Environmental migration and cities in the context of global environmental change", *Current Opinion in Environmental Sustainability*, Vol. 2 No. 3, pp. 161-165.

Adger, W.N., Arnell, N.W., Black, R., Dercon, S., Geddes, A. and Thomas, D.S. (2015), "Focus on environmental risks and migration: causes and consequences", *Environmental Research Letters*, Vol. 10 No. 6, p. 060201.

Adger, W.N., Dessai, S., Goulden, M., Hulme, M., Lorenzoni, I., Nelson, D.R., Naess, L.O., Wolf, J. and Wreford, A. (2009), "Are there social limits to adaptation to climate change?", *Climatic Change*, Vol. 93 No. 3, pp. 335-354.

Adger, W.N., Kelly, P.M., Winkels, A., Huy, L.Q. and Locke, C. (2002), "Migration, remittances, livelihood trajectories, and social resilience", *AMBIO: A Journal of the Human Environment*, Vol. 31 No. 4, pp. 358-366.

Afifi, T. (2011), "Economic or environmental migration? The push factors in Niger", *International Migration*, Vol. 49 No. s1, pp. e95-e124.

Ahmed, A.U., Hassan, S.R., Etzold, B. and Neelormi, S. (2012), *Where the Rain Falls, Case Study: Bangladesh*, United Nations University and Institute for Environment and Human Security (UNU-EHS), Bonn.

BAMETI (2017), *Agricultural Profile of the State*, BAMETI, Bihar.

Banerjee, S., Gerlitz, J.I. and Hoermann, B. (2011), "Labour migration as a response strategy to water hazards in the Hindu Kush-Himalayas. Kathmandu, Nepal", *ICIMOD*.

Barnett, J. and Webber, M. (2009), "Accommodating migration to promote adaptation to climate change", *The Commission on Climate Change and Development*, available at: www.ccdcommission.org (accessed on 5).

Below, T., Artner, A., Siebert, R. and Sieber, S. (2010), "Micro-level practices to adapt to climate change for African small-scale farmers", A Review of Selected Literature, Environ. Prod. Technol. Division, p. 953.

Bhatta, G.D. and Aggarwal, P.K. (2016), "Coping with weather adversity and adaptation to climatic variability: a cross-country study of smallholder farmers in South Asia", *Climate and Development*, Vol. 8 No. 2, pp. 145-157.

Bhatta, G.D., Aggarwal, P.K., Kristjanson, P. and Shrivastava, A.K. (2016), "Climatic and non-climatic factors influencing changing agricultural practices across different rainfall regimes in South Asia", *Current Science*, Vol. 110 No. 7, p. 1272.

Bhatta, G.D., Aggarwal, P.K., Poudel, S. and Belgrave, D.A. (2015a), "Climate-induced migration in South Asia: migration decisions and the gender dimensions of adverse climatic events", *Journal of Rural and Community Development*, Vol. 10 No. 4.

Bhatta, G.D., Aggarwal, P.K., Shrivastava, A.K. and Sproule, L. (2015b), "Is rainfall gradient a factor of livelihood diversification? Empirical evidence from around climatic hotspots in Indo-Gangetic plains", *Environment, Development and Sustainability*, Vol. 18 No. 6, pp. 1657-1678.

Boyd, R. and Ibarrarán, M.E. (2009), "Extreme climate events and adaptation: an exploratory analysis of drought in Mexico", *Environment and Development Economics*, Vol. 14 No. 3, pp. 371-395.

Brooks, N. and Adger, W.N. (2005), "Assessing and enhancing adaptive capacity", Adaptation Policy Frameworks for Climate Change: Developing Strategies, Policies and Measures, pp. 165-181.

Bryan, E., Deressa, T.T., Gbetibouo, G.A. and Ringler, C. (2009), "Adaptation to climate change in Ethiopia and South Africa: options and constraints", *Environmental Science & Policy*, Vol. 12 No. 4, pp. 413-426.

Census of India (2011), *Government of India*, New Delhi.

Climate of Bihar (2009), *Indian Metrological Department*, Pune.

Conisbee, M. and Simms, A. (2003), *Environmental Refugees: The Case for Recognition*, New Economics Foundation, London, pp. 17-18.

Conway, D. and Cohen, J.H. (1998), "Consequences of migration and remittances for Mexican transnational communities", *Economic Geography*, Vol. 74 No. 1, pp. 26-44.

Cunguara, B., Langyintuo, A. and Darnhofer, I. (2011), "The role of nonfarm income in coping with the effects of drought in southern Mozambique", *Agricultural Economics*, Vol. 42 No. 6, pp. 701-713.

Curtis, K.J. and Schneider, A. (2011), "Understanding the demographic implications of climate change: estimates of localized population predictions under future scenarios of sea-level rise", *Population and Environment*, Vol. 33 No. 1, pp. 28-54.

De Haan, A., Brock, K. and Coulibaly, N. (2002), "Migration, livelihoods and institutions: contrasting patterns of migration in Mali", *Journal of Development Studies*, Vol. 38 No. 5, pp. 37-58.

De Haas, H. (2010), "Migration and development: a theoretical perspective1", *The International Migration Review*, Vol. 44 No. 1, pp. 227-264.

Deshingkar, P. and Akter, S. (2009), *Migration and Human Development in India*, Research Paper 2009/13, Human Development Reports, UNDP.

Deshingkar, P. and Start, D. (2003), *Seasonal Migration for Livelihoods in India: Coping, Accumulation and Exclusion*, Overseas Development Institute, London.

Etzold, B., Ahmed, A.U., Hassan, S.R. and Neelormi, S. (2014), "Clouds gather in the sky, but no rain falls. vulnerability to rainfall variability and food insecurity in northern Bangladesh and its effects on migration", *Climate and Development*, Vol. 6 No. 1, pp. 18-27.

Fei, J.C.H. and Ranis, G. (1964), *Development of the Labor: Surplus Economy*, ed., Irwin.

Gbetibouo, G.A. (2009), "Understanding farmers' perceptions and adaptations to climate change and variability: the case of the Limpopo basin, South Africa", *International Food Policy Research Institute*, Vol. 849.

Gray, C. and Bilsborrow, R. (2013), "Environmental influences on human migration in rural Ecuador", *Demography*, Vol. 50 No. 4, pp. 1217-1241.

Gray, C. and Mueller, V. (2012), "Drought and population mobility in rural Ethiopia", *World Development*, Vol. 40 No. 1, pp. 134-145.

Grothmann, T. and Patt, A. (2005), "Adaptive capacity and human cognition: the process of individual adaptation to climate change", *Global Environmental Change*, Vol. 15 No. 3, pp. 199-213.

Hansen, J.W., Marx, S.M. and Weber, E.U. (2004), "The role of climate perceptions, expectations, and forecasts in farmer decision making: the argentine pampas and South Florida", *Final Report of an IRI Seed Grant Project*.

Harris, J.R. and Todaro, M.P. (1970), "Migration, unemployment and development: a two-sector analysis", *The American Economic Review*, Vol. 60 No. 1, pp. 126-142.

Hassan, R. and Nhemachena, C. (2008), "Determinants of African farmers' strategies for adapting to climate change: Multinomial choice analysis", *African Journal of Agricultural and Resource Economics*, Vol. 2 No. 1, pp. 83-104.

Henry, S., Schoumaker, B. and Beauchemin, C. (2004), "The impact of rainfall on the first out-migration: a multi-level event-history analysis in Burkina Faso", *Population and Environment*, Vol. 25 No. 5, pp. 423-460.

Hunter, L.M., Murray, S. and Riosmena, F. (2013), "Rainfall patterns and US migration from rural Mexico", *International Migration Review*, Vol. 47 No. 4, pp. 874-909.

Huntley, B. (1999), "Climatic change and reconstruction", *Journal of Quaternary Science*, Vol. 14 No. 6, pp. 513-520.

Kates, R.W. (2000), "Cautionary tales: adaptation and the global poor", *Societal Adaptation to Climate Variability and Change*, Springer, Netherlands, pp. 5-17.

Keshri, K. and Bhagat, R.B. (2013), "Socioeconomic determinants of temporary labour migration in India: a regional analysis", *Asian Population Studies*, Vol. 9 No. 2, pp. 175-195.

Klaiber, H.A. (2014), "Migration and household adaptation to climate: a review of empirical research", *Energy Economics*, Vol. 46, pp. 539-547.

Kohli, H.S., Sharma, A. and Sood, A. (Eds) (2011), *Asia 2050: Realizing the Asian Century*, SAGE Publications India.

Kumar, K.K. and Viswanathan, B. (2012), *Weather Variability and Agriculture: Implications for Long and Short-Term Migration in India*, Centre for Development Economics Department of Economics, Delhi School of Economics.

Laczko, F. and Aghazarm, C. (Eds) (2009), *Migration, Environment and Climate Change: Assessing the Evidence*, International Organization for Migration, Geneva, pp. 7-40.

Livingstone, D.N. (2000), "Environmental determination", in Johnston, R.J., Gregory, D., Pratt, G. and Watts, M. (Eds), *The Dictionary of Human Geography*, 4th ed., Blackwell, Oxford, pp. 212-215.

Lucas, R.E. (1997), "Internal migration in developing countries", *Handbook of Population and Family Economics*, Vol. 1, pp. 721-798.

Lybbert, T.J. and Sumner, D.A. (2012), "Agricultural technologies for climate change in developing countries: policy options for innovation and technology diffusion", *Food Policy*, Vol. 37 No. 1, pp. 114-123.

Maddison, D. (2006), "The perception of and adaptation to climate change in Africa", *CEEPA. Discussion Paper No. 10*, Centre for Environmental Economics and Policy in Africa. University of Pretoria, Pretoria.

Marin, A. (2010), "Riders under storms: contributions of nomadic herders' observations to analysing climate change in Mongolia", *Global Environmental Change*, Vol. 20 No. 1, pp. 162-176.

Marx, S.M., Weber, E.U., Orlove, B.S., Leiserowitz, A., Krantz, D.H., Roncoli, C. and Phillips, J. (2007), "Communication and mental processes: experiential and analytic processing of uncertain climate information", *Global Environmental Change*, Vol. 17 No. 1, pp. 47-58.

Massey, D.S., Arango, J., Hugo, G., Kouaouci, A., Pellegrino, A. and Taylor, J.E. (1993), "Theories of international migration: a review and appraisal", *Population and Development Review*, Vol. 19 No. 3, pp. 431-466.

McDowell, C. and Haan, A.D. (1997), "Migration and sustainable livelihoods: a critical review of the literature", Working Paper-Institute of Development Studies, University of Sussex.

McLeman, R. and Smit, B. (2006), "Migration as an adaptation to climate change", *Climatic Change*, Vol. 76 Nos 1/2, pp. 31-53.

Mincer, J. (1978), "Family migration decisions", *Journal of Political Economy*, Vol. 86 No. 5, pp. 749-773.

Mueller, V., Gray, C. and Kosec, K. (2014), "Heat stress increases long-term human migration in rural Pakistan", *Nature Climate Change*, Vol. 4 No. 3, pp. 182-185.

Nhemachena, C. and Hassan, R. (2007), "Micro-Level Analysis of Farmers' Adaptation to Climate Change in Southern Africa", *IFPRI Discussion Paper No. 00714*, International Food Policy Research Institute, Washington, DC.

Nordas, R. and Gleditsch, N.P. (2007), "Climate change and conflict", *Political Geography*, Vol. 26 No. 6, pp. 627-638.

Patnaik, U. and Das, P.K. (2017), "Do development interventions confer adaptive capacity? Insights from rural India", *World Development (Article in Press)*, Vol. 97, pp. 298-312.

Patnaik, U. and Narayanan, K. (2015), "How effective are coping mechanisms in securing livelihoods against climatic aberrations?: Evidences from rural India", *International Journal of Climate Change Strategies and Management*, Vol. 7 No. 3, pp. 359-374.

Porter, J.R., Xie, L., Challinor, A.J., Cochrane, K., Howden, S.M., Iqbal, M.M., Lobell, D.B. and Travasso, M.I. (2014), *Chapter 7: Food Security and Food Production Systems*, Cambridge University Press.

Ravuvu, A. (1992), "Security and confidence as basic factor in pacific islander's migration", *The Journal of the Polynesian Society*, Vol. 101 No. 4, pp. 329-341.

Renaud, F.G., Bogardi, J.J., Dun, O. and Warner, K. (2007), *Control, Adapt or Flee: How to Face Environmental Migration?*, UNU-EHS.

Rivera, W.M. and Sulaiman, V.R. (2009), "Extension: object of reform, engine for innovation", *Outlook on Agriculture*, Vol. 38 No. 3, pp. 267-273.

Roncoli, C., Ingram, K. and Kirshen, P. (2002), "Reading the rains: local knowledge and rainfall forecasting in Burkina Faso", *Society &Natural Resources*, Vol. 15 No. 5, pp. 409-427.

Scheffran, J., Marmer, E. and Sow, P. (2012), "Migration as a contribution to resilience and innovation in climate adaptation: social networks and co-development in Northwest Africa", *Applied Geography*, Vol. 33, pp. 119-127.

Stal, M. (2011), "Flooding and relocation: the Zambezi river valley in Mozambique", *International Migration*, Vol. 49 No. s1, pp. e125-e145.

Stark, O. (1984), "Rural-to-urban migration in LDCs: a relative deprivation approach", *Economic Development and Cultural Change*, Vol. 32 No. 3, pp. 475-486.

Stark, O. and Bloom, D.E. (1985), "The new economics of labor migration", *The American Economic Review*, Vol. 75 No. 2, pp. 173-178.

Stark, O. and Levhari, D. (1982), "On migration and risk in LDCs", *Economic Development and Cultural Change*, Vol. 31 No. 1, pp. 191-196.

Stark, O. and Lucas, R.E. (1988), "Migration, remittances, and the family", *Economic Development and Cultural Change*, Vol. 36 No. 3, pp. 465-481.

Stern, N.H. (2007), *The Economics of Climate Change: The Stern Review*, Cambridge University press.

Stojanov, R. (2004), "Environmental migration: how can it be estimated and predicted?", *Geographica*, Vol. 38, pp. 77-84.

Stojanov, R., Kelman, I., Ullah, A.K.M., Duží, B., Procházka, D. and Blahůtová, K.K. (2016a), "Local expert perceptions of migration as a climate change adaptation in Bangladesh", *Sustainability*, Vol. 8 No. 12, p. 1223.

Stojanov, R., Duží, B., Kelman, I., Němec, D. and Procházka, D. (2016b), "Local perceptions of climate change impacts and migration patterns in Malé, Maldives", *The Geographical Journal*, doi: 10.1111/geoj.12177.

Swanson, B.E. (2006), "Extension strategies for poverty alleviation: Lessons from china and India", *Journal of Agricultural Education and Extension*, Vol. 12 No. 4, pp. 285-299.

Tacoli, C. (2009), "Crisis or adaptation? Migration and climate change in a context of high mobility", *Environment and Urbanization*, Vol. 21 No. 2, pp. 513-525.

Tarleton, M. and Ramsey, D. (2008), "Farm-level adaptation to multiple risks: climate change and other concerns", *Journal of Rural and Community Development*, Vol. 3 No. 2.

Taylor, E.J. (1999), "The new economics of labour migration and the role of remittances in the migration process", *International Migration (Geneva, Switzerland)*, Vol. 37 No. 1, pp. 63-88.

The Economic Survey of India (2016-17), *Ministry of Finance*, Government of India.

Tripathi, A. (2017), "Socioeconomic backwardness and vulnerability to climate change: evidence from Uttar Pradesh state in India", *Journal of Environmental Planning and Management*, Vol. 60 No. 2, pp. 328-350.

Tschakert, P. and Tutu, R. (2010), "Solastalgia: environmentally induced distress and migration among Africa's poor due to climate change", *Environment, Forced Migration and Social Vulnerability*, Springer Berlin Heidelberg, pp. 57-69.

Tumbe, C. (2011), "Remittances in India: facts & issues", *Indian Journal of Labour Economics*, Vol. 54 No. 3, pp. 479-501.

Tyson, P.D., Lee-Thorp, J., Holmgren, K. and Thackeray, J.F. (2002), "Changing gradients of climate change in southern Africa during the past millennium: implications for population movements", *Climatic Change*, Vol. 52 No. 1, pp. 129-135.

Upadhyay, H., Kelman, I., Lingaraj, G.J., Mishra, A., Shreve, C. and Stojanov, R. (2015), "Conceptualizing and contextualizing research and policy for links between climate change and migration", *International Journal of Climate Change Strategies and Management*, Vol. 7 No. 3, pp. 394-417.

Viswanathan, B. and Kumar, K.K. (2015), "Weather, agriculture and rural migration: evidence from state and district level migration in India", *Environment and Development Economics*, Vol. 20 No. 4, pp. 469-492.

Warner, K. (2010), "Global environmental change and migration: governance challenges", *Global Environmental Change*, Vol. 20 No. 3, pp. 402-413.

Warner, K. and Afifi, T. (2014), "Where the rain falls: evidence from 8 countries on how vulnerable households use migration to manage the risk of rainfall variability and food insecurity", *Climate and Development*, Vol. 6 No. 1, pp. 1-17.

West, C.T., Roncoli, C. and Ouattar, F. (2008), "Local perceptions and regional climate trends on central plateau of Burkina Faso", *Land Degradation & Development*, Vol. 19 No. 3, pp. 289-304.

West, C.T., Vásquez-Léon, M. and Finan, T.J. (2003), 12.3 Testing Farmers' perceptions of Climate Variability: A Case Study From The Sulphur Springs Valley, AZ.

Woodruff, C. and Zenteno, R. (2007), "Migration networks and microenterprises in Mexico", *Journal of Development Economics*, Vol. 82 No. 2, pp. 509-528.

Yi, S., Chengjin, X., Hailing, Z. and Zheng, W. (2017), "Migration in response to climate change and its impact in china", *International Journal of Climate Change Strategies and Management*, Vol. 9 No. 3.

Further reading

Bedford, R., Ho, E. and Lidgard, J. (2000), *International Migration in New Zealand: Context, Components and Policy Issues*, University of Waikato, Population Studies Centre.

About the authors

Chandan Kumar Jha is a Fellow (PhD) Student with National Institute of Industrial Engineering, Mumbai, India. He holds a masters' degree in Economics. His research interests include Climate change and agriculture, Adaptation strategies to increase crop productivity and crop income. Chandan Kumar Jha is the corresponding author and can be contacted at: chandan1929@gmail.com

Vijaya Gupta has been in teaching and research for the past 30 years. She is currently working as a Professor in Economics at the National Institute of Industrial Engineering, Mumbai. Her research interest is in the areas of environmental and agricultural economics in general and valuation of intangible, vulnerability assessment due to climate change, growth and environment, inequality in particular.

Dr Utpal Chattopadhyay is currently working as an Associate Professor, Economics & Strategy Area at National Institute of Industrial Engineering (NITIE), Mumbai, India. Earlier, he had worked for more than a decade in management training and consultancy domain with the National Productivity Council of India (NPC). He also had a short stint of working in a research project at the Indian School of Mines (ISM), Dhanbad. Dr Chattopadhyay holds PhD degree in Economics from Delhi University and MSc (Economics) from Calcutta University. His teaching and research interests include areas such as managerial economics, economic environment of business, global competitiveness and international trade issues.

Dr Binilkumar Amarayil Sreeraman is an Assistant Professor of Economics in National Institute of Industrial Engineering (NITIE), Mumbai, India. He teaches courses on Managerial Economics, Macro Economics, Resource Economics and Industrial Organization for PG Students. He has received his Doctoral Degree in Economics from Indian Institute of Technology Bombay in 2010. His broad research interests lie in Environmental and Natural Resource Economics, Developmental Economics, Industrial Organization, Banking and Environmental Policy. He has published research articles in refereed national and international journals and also has presented research papers at refereed international conferences. He is professionally affiliated with the academic bodies such as International Society of Ecological Economics (ISEE), European Association of Environmental and Resource Economists (EAERE), Global Studies Research Network, Indian Society of Ecological Economics (INSEE), Indian Society of Agricultural Economics, World Wide Fund for Nature India (WWF India), etc.

7

Increasing vulnerability to floods in new development areas: evidence from Ho Chi Minh City

Phan N. Duy
Department of Urban Planning, University of Architecture of Ho Chi Minh City, Ho Chi Minh City, Vietnam and School of Geography, Earth and Environment Sciences, University of Birmingham, Birmingham, UK

Lee Chapman
School of Geography, Earth and Environment Sciences, University of Birmingham, Birmingham, UK

Miles Tight
School of Civil Engineering, University of Birmingham, Birmingham, UK

Phan N. Linh
Department of Urban Planning, University of Architecture of Ho Chi Minh City, Ho Chi Minh City, Vietnam, and

Le V. Thuong
Southern Institute for Spatial Planning of Vietnam, Ho Chi Minh City, Vietnam

Abstract

Purpose – Flooding is an emerging problem in Ho Chi Minh City (HCMC), Vietnam, and is fast becoming a major barrier to its ongoing development. While flooding is presently of nuisance value, there is a growing concern that a combination of rapid urban expansion and climate changes will significantly exacerbate the problem. There has been a trend of population being rapidly accommodated in new urban areas, which are considered highly vulnerable to floods, while the development strategy by the local government still attracts more property investments into the three new districts on the right side of Saigon River. This paper aims to discuss the increase in the number of residences vulnerable to flooding, to underline the need for more appropriate future spatial development. For the vision, an application of compact and resilient theories to strategic planning and management of this city is proposed to reduce vulnerability. This paper also highlights the need to better understand growing vulnerability to floods related to urban expansion over low-lying former wetlands and the more important role of planning spatial development accompanied with transportation investment which can contribute to flooding resilience.

Design/methodology/approach – This research uses combined-methods geographical information system (GIS) analysis based on secondary data of flood records, population distributions, property development (with the details of 270 housing projects compiled as part of this research) and flooding simulation. This allows an integrated approach to the theories of urban resilience and compactness to discuss the implication of spatial planning and management in relevance to flooding vulnerability.

Findings – The flooding situation in HCMC is an evidence of inappropriate urban expansion leading to increase in flooding vulnerability. Although climate change impacts are obvious, the rapid population growth and associated accommodation development are believed to be the key cause which has not been solved. It was found that the three new emerging districts (District 2, 9 and ThuDuc) are highly vulnerable to floods, but the local government still implements the plan for attracted investments in housing without an integrated flooding management. This is also in line with the development pattern of many coastal cities in Southeast Asia, as economic development can be seen as a driving factor.

Research limitations/implications – The data of property development are diversified from different sources which have been compiled by this research from the basic map of housing investments from a governmental body, the Department of Construction. The number of projects was limited to 270 per over 500 projects, but this still sufficiently supports the evidence of increasing accommodation in new development districts.

Practical implications – HCMC needs neater strategies for planning and management of spatial development to minimize the areas vulnerable to floods: creating more compact spaces in the central areas (Zone 1) protected by the current flooding management system, and offering more resilient spaces for new development areas (Zone 2), by improving the resilience of transportation system. Nevertheless, a similar combination of compact spaces and resilient spaces in emerging districts could also be incorporated into the existing developments, and sustainable drainage systems or underground water storage in buildings could also be included in the design to compensate for the former wetlands lost.

Social implications – This paper highlights the need to better understand growing vulnerability to floods related to urban expansion over low-lying former wetlands and emphasizes the more important role of planning spatial development accompanied with transportation investment which can contribute to flooding resilience. Coastal cities in southeast countries need to utilize the former-land, whereas feasibility of new land for urban expansion needs to be thoroughly considered under risk of natural disasters.

Originality/value – A combination of compact spaces with improved urban resilience is an alternative approach to decrease the flooding risk beyond that of traditional resistant systems and underlines the increasingly important role of urban planning and management to combat the future impacts of floods.

Keywords Resilience, Compact city, Flooding vulnerability, Spatial planning and management, Urban expansion

1. Introduction

Over 50 per cent of the world's population now lives in urban areas (United Nations – UN, 2014). Growth can be traced back to the nineteenth century where economic drivers provided an impulse for urban expansion as a direct impact of industrialism (Hall and Jones, 2011): a trend that has continued to the present day. In 2010, 167 cities across the world had over 7,50,000 citizens, a result of a 10- to 20-fold increase in population since 1960 (United Nations – UN, 2012). However, as cities continue to grow, the development pressures of new urban areas create an increased concentration of people and assets with a greater potential for devastation from natural disasters. A growing imbalance between the natural and human environment has led to higher risks of floods in major cities (Cigler, 2007). This problem appears to be particularly acute in low gross domestic product countries, as many cities have experienced uncontrolled urban enlargement on low-lying lands. Based on an investigation of 90 cities worldwide, Angel *et al.* (2005) projected that the built-up areas will increase from 2,00,000 km^2 in 2010 to 6,00,000 km^2 by 2030 in developing-country cities. Dasgupta *et al.* (2012) stated that coastal wetlands, which are natural barriers to coastal flooding, have been gradually lost to the process of urban expansion in 76 developing countries, of which "Vietnam is by far the most vulnerable country with close to 65 per cent of its freshwater marshes at risk". Linked to this, Dutta (2011) emphasizes the abnormal frequency of flood occurrence in Asian countries, where several cities have faced rapid expansion but owing to a changing climate are now at higher risks of flooding incidents (examples include Bangkok, Manila and Ho Chi Minh City: Asia Development Bank – ADB, 2010).

Ho Chi Minh City (HCMC), a coastal city of Vietnam, is considered to be particularly vulnerable to climate change (World Bank – WB, 2010). It is one of 20 port cities worldwide that have been identified as having both a high rate of population growth and a large number of assets projected to be exposed to sea level rise by 2070 (Nicholls, 2011). In parallel with economic development, despite progressive investments into flood protection system, the frequency and severity of floods have become more significant. The city actually rarely experienced floods until the 1960s (Hong, 2011), but now more than 50 per cent of the urban area of HCMC is affected by regular floods (Asia Development Bank – ADB, 2010). From 2010 to 2015, the Steering Center of the Urban Flood Control Program of HCMC (SCFC) has catalogued 980 inundations which not only endanger citizens but also act as a barrier to the economic prospects of the city. While present flooding in HCMC is frequent, it is presently more of nuisance value causing low level, but widespread, disruption; however, there are concerns that HCMC could begin to experience "extreme" flooding, such as that seen in Bangkok in 2011, if there remains a lack of an effective planning strategy for the urban spatial enlargement (Phi, 2013). Future scenarios indicate that flood-prone areas in HCMC are projected to increase and cover 49 per cent of the built-up area by 2025, if the 2.0 m water level is reached (Storch and Downes, 2011), and to 71 per cent by 2050 in case of a combined flood (Asia Development Bank – ADB, 2010). These have resulted in divergent opinions on the link between flooding situation and urban planning and management. Given these views and the rapid expansion of HCMC, this paper investigates flooding vulnerability in relation to the spatial enlargement, particularly in new development areas close to Saigon River.

2. Ho Chi Minh City: a coastal city increasingly vulnerable to floods

As an emerging coastal mega-city, the population of HCMC increased from 3.5 million in 1976 to 8.2 million in 2015 (General Statistic Office of Vietnam – GSOV, 2015), but the actual number of people (including unregistered migrants) is now estimated to be nearly 10 million (Asia Development Bank – ADB, 2010). With the current territory of 2,095 km^2, the city consists of three zones (Figure 1):

(1) *Zone 1-13* central districts including the "old-town" (partly located in District 1 and 5);

(2) *Zone 2-6* new development districts where there have been many new residential developments; and

(3) *Zone 3-5* suburban districts which used to be the rural land in the early stage.

The growth of HCMC can be attributed to its location as a port city on river basin, attracting a large amount of labor as the driving factor for economic development, but this now provides some disadvantages related to the characteristics of geography and topography. Located 50 km from the Southeast Asia sea, it is considered one of the three "hot-spots" in this region which is vulnerable to climate changes such as sea level rise (World Bank – WB, 2010; Asia Development Bank – ADB, 2010). Owing to the location on a low-lying coastal plain, 40-45 per cent of land of HCMC is elevated from 0 to 1 m from sea level (Asia Development Bank – ADB, 2010) and more than 50 per cent of land of HCMC is under 1.5 m from sea level (Thinh *et al.*, 2009). Along with a process of spatial expansion, the old-town (the core-part of District 1, Zone 1) is found on relatively high-land, but almost all new giant

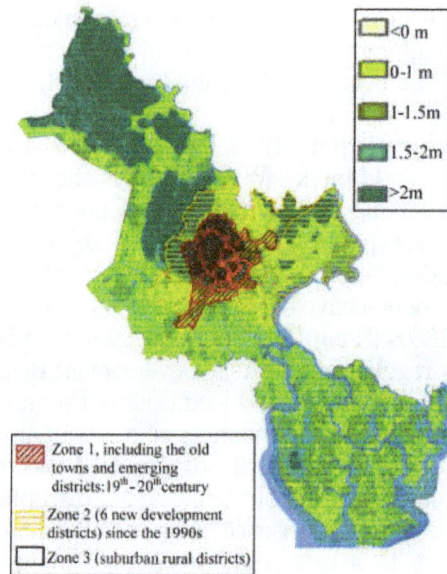

Figure 1. Urban zones layered over topography

developments in Zone 2, such as PhuMyHung (District 7) and ThuThiem (District 2), are built on areas with a higher risk of flooding owing to the low-lying nature of peripheral areas earmarked for development. The vulnerability of flooding in HCMC is assessed in this paper by using a geographical information system (GIS) approach. The analysis contains population changes, land occupation for property investments and flood impact simulations to ascertain to what extent urban development is leading to an increase in vulnerability to floods in new development areas. Based on this analysis, an extrapolation of the trend of urban settlement over vulnerable areas to floods, especially in three emerging districts on the eastern side of Saigon River, is needed to examine the future implications for the city. Finally, through the vision for HCMC, the paper will discuss issues related to a potential link between the theories of compact city and resilient city to propose a strategy of urban planning and management as the long-term solution to flooding.

2.1. Urban expansion and flooding impacts

Since the early beginnings in the seventeenth century, the original town covered about 5 km^2 with a population of around 2,00,000 people, the first rampart named GiaDinh had been built on (locally) a high-land (4-6 m), presently located in District 1. In the nineteenth century, Saigon, the old name of HCMC, was established as the most modern city in Southeast Asia in the light of a development plan to expand the city to 25 km^2, with a view to accommodating 5,00,000 people. Until the end of twentieth century, the built-up area was limited to the left of Saigon River.

By the beginning of the twenty-first century, widespread and uncontrolled urban enlargement of HCMC was occurring on previously rural land. Since the establishment of six new districts in 1997, there was a shifting trend of dwellers moving from the old center to the new suburbs, but such changes in population distribution had not adequately considered the vulnerability to floods (Storch and Downes, 2011). In fact many of the new development areas close to Saigon River were vulnerable to fluvial floods; Thinh *et al.* (2009) reported five districts affected by tides in combination with heavy rains in 2008. Some of these, such as BinhThanh and District 2 in Zone 2, were more susceptible to floods because of the higher rate of urbanization resulting in high-density residences (Tu, 2010). In the 2000s, this is also evident with the high-rate of construction across emerging districts (e.g. 2, 6, 7, 8, 12 and Binh-Tan from 1989 to 2002: Viet, 2008). An illustration of this is the rapid development of PhuMyHung (409 ha, in District 7) and ThuThiem (657 ha, in District 2), which are the two ignition projects of a larger process containing a vast number of property developments. These projects not only trigger the enlargement of urban areas accompanied with higher concentration of people and assets but also negatively affect indigenous hydrological systems, including green spaces and water bodies. Indeed, there was a rise in the area of impermeable surface versus a decrease in areas of water from 1985 to 2010 (Phi, 2013) [Figure 2(a)], and it appears that the link between new development and flood frequency is striking.

On the other hand, HCMC has also been exposed to noticeable climatic changes related to precipitation regimes and water levels in Saigon River. For example, there is a trend of a higher frequency of heavy rains (more than 50 mm) from 1999 to 2009 (Phi, 2013). However, upstream rainfall is only an exacerbating factor to flooding instead fluvial floods from the intense monsoon rainfall is more significant (Asia Development Bank – ADB, 2010). The observation data showed the increase in water levels at the PhuAn station (within Saigon River), where the levels were higher than those at the VungTau station (in the sea mouth) from 2010 to 2015 (SRHC, 2010-2015). This implies that the riverine level of HCMC has increased not only by global (sea level rise) SLR but also by other reasons related to urbanization during development such as reduction in area of agriculture and water and land subsidence by high density of construction (Phi, 2013):

(1)   Overall, the footprint of the city has rapidly changed since the 1990s with the presence of new development districts and "rural districts" in peripheral areas (particularly to the east and southeast). The development process has seen the city extend its boundary across the main water bodies, Saigon River in the East and ThiNghe-Te channel in the South to accommodate the rapid population growth. Consequently, there has been depletion in area of wetlands which are essential for natural protection for the city. These lands had previously acted as a natural buffer with a large network of water bodies, including rivers, canals and channels, but now hard surfaces are further influencing resilience of the broader city to flooding. While climate changes such as SLR and uncertain precipitation are general factors in HCMC, uncontrolled urbanization is believed to be the dominant factor for increased flooding in HCMC (Hong, 2011; Storch and Downes, 2011; Phi, 2013). This is a challenge for flood management because of dynamic social changes in combination with the uncertain nature of climatic changes.

2.2. Current flood management
The master plan for water discharge toward 2020 (since decision 752 by the central government in 2001) identified six zones consisting of the current inner city (about 140 km^2) and surrounding areas (about 510 km^2) from which water would be discharged through dedicated channels (Figure 3). A supplementary plan with an expanded area for management (decision 1547) was actioned in 2008 to optimize the water storage capacity of current water bodies of HCMC and to control flash floods from upper regions and upstream water from the main rivers (ibid). According to this plan, the water level of +1.32 m, has been chosen for following designs, but the highest tides

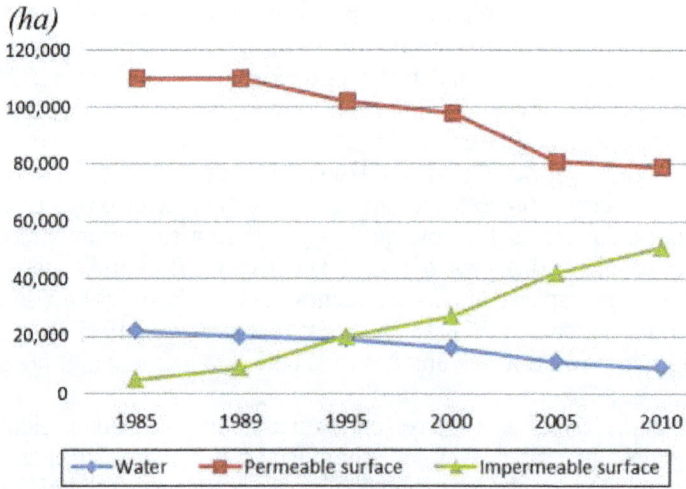

(a)

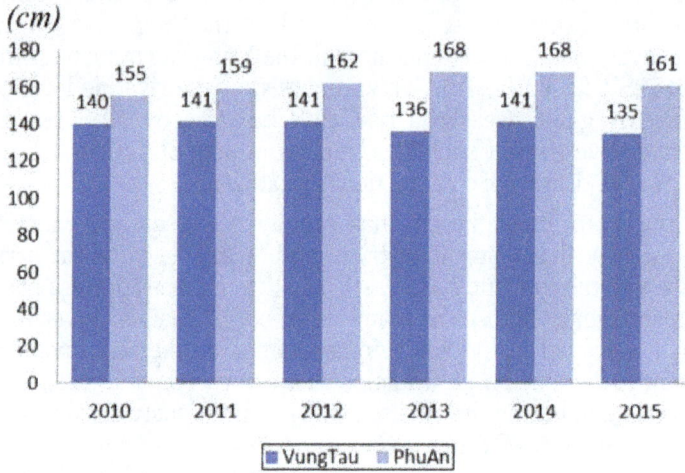

(b)

Notes: (a) Changes in areas of water, permeable surfaces and impermeable surfaces since population growth (Phi, 2013); (b) water levels in comparison between VungTau (East Sea) and PhuAn (SaiGon River) SRHC (2010-2015)

Figure 2. Changing factors related to floods

were recently recorded at +1.68 m in 2013 and 2014 (Southern Regional Hydro – Meteorological Center – SRHC, 2015). Fundamentally, these plans focus on utilization of existing water bodies. The flooding control system mainly consists of sluices and embankment from which the safety range to current thresholds is likely insufficient; this system, once completed by 2020, will not cover the new districts on the right side of Saigon River (Figure 3).

Urgent actions are now required, the city has been accelerating existing projects such as dredging channels, improving drainage systems, building an embankment along the left side of

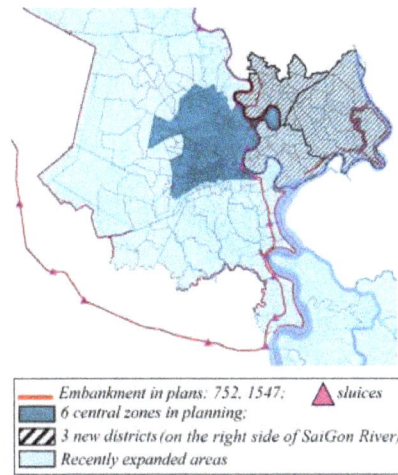

Figure 3. Flooding management in zones

Saigon River, constructing of sluices and creation of reservoirs. However, their effectiveness is in question, despite a cost of around US$1bn, floods still frequently happen in this city. It is argued that the current proposals are just short-term solutions based on engineering structures while the city has not resolved the poor spatial planning (Phuc, in Nguyen, 2015). Long-term flood management ultimately needs to control the rapid urbanization process, especially in new development areas (Storch *et al.*, 2009; Storch and Downes, 2011; Hong, 2011; Phi, 2013). Hence, this paper examines to what extent such process increases flooding vulnerabilities of HCMC, especially in the developing districts on the eastern side of Saigon River.

3. Data and methodology

Risks from climate change impacts arise from the interaction between hazard (triggered by an event or trend related to climate change), vulnerability (susceptibility to harm) and exposure (people, assets or ecosystems at risk) (International Panel on Climate Change – IPCC, 2014, p. 36).

In this research, the main method is GIS analysis which is based on secondary data of flood records, population distributions, property development (with the details of 270 housing projects compiled as part of this research) and flooding simulation. These data were made available from institutional bodies in Vietnam and other agencies and are summarized in Table I.

Table I. Key datasets used for analysis

Type of data	Available period	Source
Population distribution	1981-2015	General Statistic Office of Vietnam (GSOV)
Property development	2010-2015	Information collected by this research, relied on the property investment map from Department of Construction of HCMC (COD)
Flooding event record	2010-2015	Steering Center of Flood Control Program of Ho Chi Minh City (SCFC)
Flooding simulation	2010-2050	Asia Development Bank – ADB (2010), with the maps of flooding prepared by ICEM (2009)

From this analysis, current and future vulnerability can be measured. Vulnerability is defined as the extension of harm which is evaluated by the implications of exposure, susceptibility and resilience under particular conditions (Balica and Wright, 2009; Hufschmidt, 2011; Scheuer *et al.*, 2010; Willroth *et al.*, 2010; Fuchs *et al.*, 2011, cited in Balica *et al.*, 2012, p. 75). There is a need to drill down into the different components of vulnerability (Balica *et al.*, 2012):

- *exposure* – location of people and other assets in relation to flood prone areas;
- *susceptibility* – factors influencing degree of impacts such as awareness and preparation of citizens (aided by institutional response); and
- *resilience* – ability of a city to adapt to floods by re-organizing itself to maintain acceptable levels of function and structure.

In the case of HCMC, flooding vulnerability is defined as the probability of flood impacts in a particular area. Exposure (i.e. increased concentration of people in flood prone areas) is the core factor but susceptibility and resilience can significantly increase or reduce the degree of vulnerability. The susceptibility depends on location and time where and when a flood occurs, whereas resilient capacity can be evaluated through the possibility of evacuation and recovery through the transportation system in case of an extreme flood.

4. Assessing vulnerability to floods

4.1 Exposure

Alongside with urban expansion, the population growth rate in Zone 2 (6 new development districts) has been extremely high, particularly notable from 2003 onward in contrast to Zone 1 (13 central districts) [Figure 4(a)]. The data of property development showed that the area of land for the housing projects was 725 ha for implementation from 2010 to 2015, and it is predicted to increase by 3,670 ha in 2030. In comparison with Zone 1, such land area was much higher in Zone 2 [Figure 4(b)]. The subsequent GIS analysis (overlaying land occupied by these projects with areas affected by floods) indicates a majority of residential developments located in vulnerable area [Figure 5(a)], and this trend will continue by 2050 in reference to the extreme flooding scenario [Figure 5(b)]. Such developments are now starting to include more peripheral areas, but these areas are of low elevation and particularly prone to fluvial flooding. Indeed, the record of floods by SCFC supported that there was a rising proportion of number and length of inundation, especially in Zone 2, from 29 per cent in 2010 to 52 per cent in 2015 [Figure 4(c)]. Obviously, there have been an increase in the number of people and their assets exposed to flooding.

4.2 Susceptibility

Owing to the characteristics of the tropical weather experienced and its location in the Mekong delta region, HCMC often has high precipitation in annual rainy season (mainly from May to November), while it is also influenced by "semi-diurnal tides" (mainly from September to December), two nearly equal high and low tides a day with a shifting time of about 50 min on each contiguous day (TEDISouth, 2010). Hence the city is highly susceptible to floods from September to October, when there are more potentially combined effects with heavy rains. The statistics of flooding events by months (SCFC, 2010-2015) support that the highest level and frequency of floods is normally from September to November every year. Furthermore, the impacts are more disruptive if floods happen at peak times (6-8 a.m.; 4-6 p.m.) when urban activities are at their daily peak through physical transportation (mainly on roads). For example, the extreme flood on September 15, 2015, occurred after a heavy and long rain being

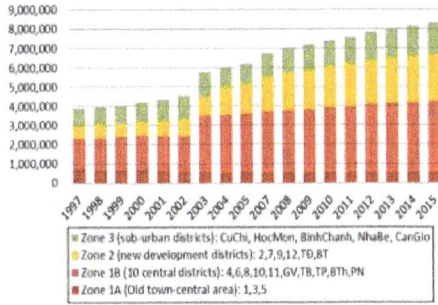

(a)

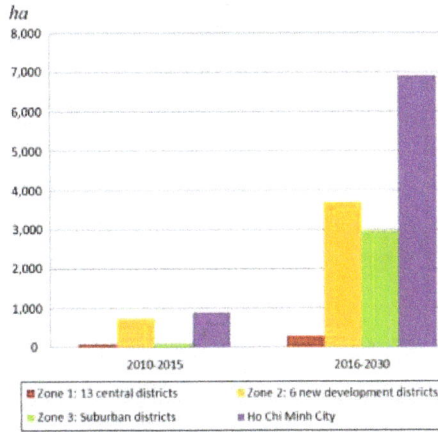

(b)

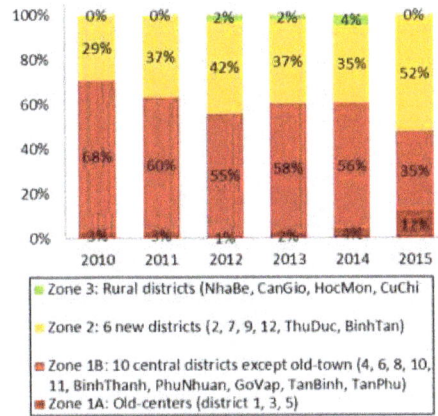

(c)

Notes: (a) Population distribution GSOV (1997-2015); (b) land for housing development from 2010-2030; (c) length and location of roads affected by floods SCFC (2010-2015)

Figure 4. Urban development and flood impacts

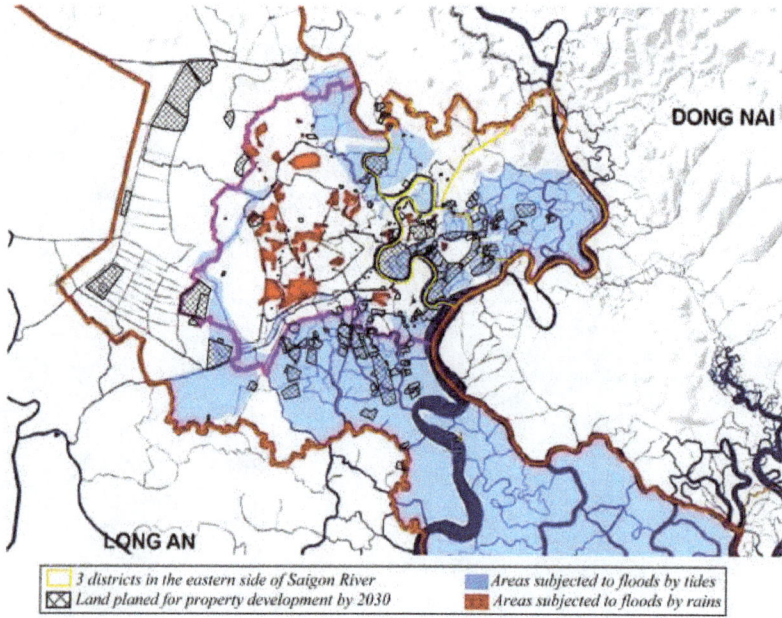

(a)

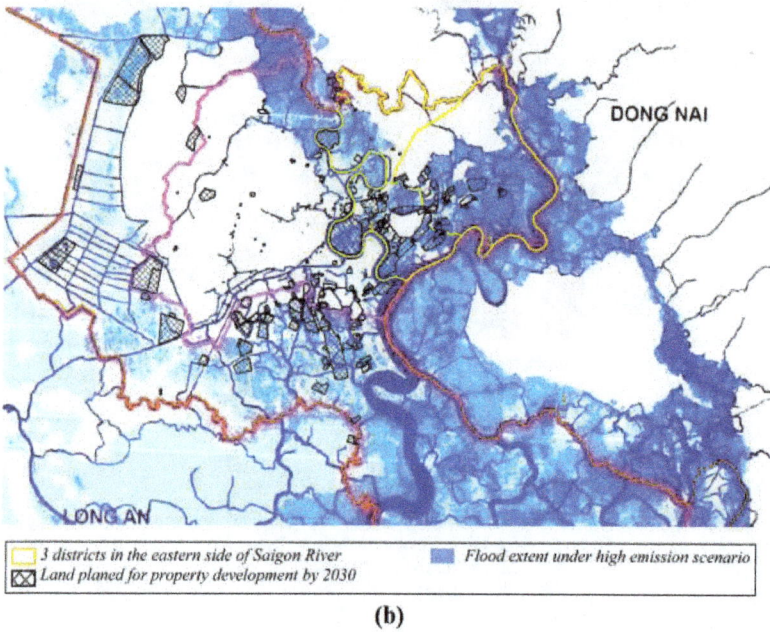

(b)

Notes: (a) Properties exposed to regular floods; flood map adapted from ICEM (2009); (b) properties exposed to extreme floods under high emission scenario; flood map adapted from ICEM (2009)

Figure 5. Property developments exposed to floods

followed by a high tide around 4-6 p.m., resulting in a large-scale effect to urban transportation. In brief, high tides can be seen as the "background/regular factor", heavy rains are the "exacerbated/uncertain factor"; and uncontrolled housing developments is actually the driving factor for increasing people and assets exposed to floods.

4.3 Resilience

Urban resilience can help the city to return to normal situation after flooding effects. However, the dependency on flooding control not only influences riverine ecosystems but also increases long-term flood risk (Burby *et al.*, 2000; Smits *et al.*, 2006), and this is not a reliable approach to climatic uncertainties (Zevenbergen and Gersonius, 2007). On the other side, urban resilience can be undermined if natural effects create an obstruction to transportation system because transport network has a vital role in connection between urban physical structures and in maintenance of urban operation. Desouza and Flanery (2013) stated that it is a physical element contributing to key urban resources for resilience. The GIS analysis by this research explores that a large-scale flood, such as the extreme incident in Bangkok, will affect to vulnerable sessions of several roads in HCMC and can divide the city into two parts: the western and the eastern side of Saigon River, regardless of the performance of embankment flood defenses (Figure 3). In particular, this causes an isolation of three development districts (2, 9 and ThuDuc), which have not been covered by the current protection system. As a result, there will be a large number of new residences isolated by floods in these places to which the arteries from the city center are potentially inaccessible. Referring to the high flood on September 15, 2015, many main streets have been obstructed up to 4 h as the vehicle flows were decelerated before blocked at important nodes such as HangXanh; as a result, many people could not return to their homes in these districts. This provides an example of how the resilience of this city is compromised by floods affecting the transportation network, especially to main routes or nodes/intersections.

By using this approach, the impact of flooding can be analyzed. In the case of HCMC, the highest risk is the combination of inner floods by heavy rains with tidal floods. The flooding vulnerability is generated by the increase in number of people and assets exposed to floods in new development districts, Zone 2. This paper highlights the areas of recent property developments in Districts 2, 9 and ThuDuc, which are believed to be the most vulnerable districts to flooding effects. This requires an improved consideration of resilience for the new residences in these areas rather than just short-term solutions based on traditional flood defense.

5. Discussion

5.1 Impact of future changes (climatic and social) in Ho Chi Minh City

In line with climate change impacts related to SLR and uncertain precipitation, HCMC has continually been improving its flooding management, but the continuing mismatch between the current planning, implementation and the volatility of climatic and social factors remains considerable. Several flood defense projects, mainly deployed in central areas, are still in progress but the built-up area has rapidly extended beyond the zones identified for protection. Furthermore, there are concerns that the design and construction of city-scale protection systems are obsolete because of the many changes in precipitation and river levels. An illustration of this is the proposed elevation of the river bank system (+2.5 m, according to the plan 1547 mentioned in Section 3.2): when completed, it will be just 82 cm over the recently maximum tides at +1.68 m recorded in 2013 and 2014, but the SLR to Vietnam is projected to increase by 75-100 cm by 2100 (Asia Development Bank – ADB, 2010; Ministry of Natural Resources and Environment of Vietnam – MNRE, 2012), with the increasing rate of water level at rivers in HCMC will be higher as compared to the sea level (Phi, 2013). According to the hydrological report by the Southern Regional Hydro-meteorological Center of Vietnam–SRHC, after 34 years, the average increase in maximum water at VamKenh station (at VungTau estuary) versus PhuAn station (on Saigon River) is 13.2 and 37.8 cm, respectively (the five years period 1981-1985 compared to 2011-2015). Besides, the extreme weather, such as heavy rains, has been challenging the existing capacity of the urban infrastructure. In 2016, there was a rainfall peak at 208 mm (during 2 h), whereas the current drainage system has been designed for rains under 100 mm (Steering Center of Flood Control Program of Ho Chi Minh City – SCFC, 2016). These imply that the current flood management plans are lagging behind the changes of both climatic and human factors. It is impractical to deliver a large budget to new defences which mainly protect the central areas, but the majority of citizens have been highly accommodating in new developing areas. In the future, the threshold of defenses to floods, as the safety range, will be exceeded by climate changes, while the city has been experiencing more new residences on flood plain unless appropriate adjustments to spatial development. Fundamentally, this paper criticizes the fact that uncontrolled and subjective plan for urbanization depended on a hard-engineered protection system. This, when coupled with climate changes, will ensure the city is increasingly vulnerable to floods and is in urgent need of a long-term and comprehensive strategy, from theory to practice.

5.2 Visions for Ho Chi Minh City: resilience and compactness
Like many cities across the world, the original urban areas of HCMC have a situational advantage, e.g. higher elevation, that has traditionally made them resistant to floods, but the vulnerability increases as more and more peripheral lands are built-up. The speed of the development is crucial here. Flood management policy can be slow, and there is often long duration required to get a protection system from planning to implementation, whereas urbanization fundamentally follows the free-market of property development. This challenges any urban physical infrastructure dealing with uncertainties while hard engineered solutions can often be insufficient.

Assuming urban development is inevitable, increasing compactness in terms of spatial planning and management is potentially the answer for a city to reduce vulnerability to floods. A compact city, which adheres to the desirability of sustainable development advocated by the European Community since the 1990s (Commission of the European Community-CEC, 1990), is still contemporarily supported by scholars who emphasize the energy efficiency with the promotion of "intensification of the use of space in the city" (Elkin *et al.*, 1991, p. 16), viability of amenity and facility for social sustainability (Haughton and Hunter, 1994) and "offer the maximum scope for effective traffic reduction" (McLaren, 1992, p. 281). In several cities in Organisation for Economic Co-operation and Development (OECD) countries, its potential success has been demonstrated in meeting urban needs of public transport and accessibility to local services, while lessening environmental impacts (Matsumoto, 2012). More broadly, this paper focuses on reducing impacts in urban areas prone to floods, by considering the role of the vertical dimension as opposed to just centripetal urbanization leading to widespread construction on floodplains. Centralized, compact spaces could be deployed from urban to building scale to urban scale. "Urban flood risks should be proactively managed through resilience, taking advantage of interventions at different spatial levels" (Zevenbergen *et al.*, 2008). Subsequently, urban public space should be improved and opened for "absorbable surface" such as green areas, and there is a need of maximized function within buildings in terms of architectural design, e.g. optimized height, interconnected corridors, storage spaces for water. Transportation structures are not only the interlinks between civil buildings and urban spaces but also an incentive for improving adaptation to regular floods and a supporting element for resilience to extreme floods.

Despite the claimed advantages, the compactness theory has been debated in relevance to the different context of application (e.g. Europe, UK, US or AU) and the quality of living environment. Decentralization has remained the trend in urban development (Breheny and Rookwood, 1993), while centralization (compactness) is a choice for cities targeting sustainable development (Thomas and Cousin, 1996). As a balance, a compromising stance reconciles sustainability with controlled development (Jenks *et al.*, 1996). For example, urban decentralized areas can contain several compacted settlements interlinked by public transport (Haughton and Hunter, 1994), walking and cycling are appropriate to encourage social interaction between different residences (Elkin *et al.*, 1991) and health benefits and lower congestion (Miles *et al.*, in Chapman and Ryley, 2012). Additionally, these routes can become emergency evacuation routes in the case of an extreme flood if incorporated into defense schemes. Hence, new developments with compact spaces on advantage lands (e.g. high elevation, distance construction from flooding risk factors) can be both less exposed to floods as a promising solution to enhance urban resilience.

In the case of HCMC, buildings should be built in old-town areas to allow intensification in those areas protected by current flooding defense, whereas the new development areas should be designed such that there are more open spaces for water absorption. In coordination, "vertical transportation" in complex buildings will be an appropriate solution to uncertain floods, whereas "horizontal transportation" will ensure the contiguous connections between different concentrated areas. On the other hand, once commuting distance between different urban areas can be reduced, it is more possible for emergency evacuation through a resilient transport network, including routes elevated from the ground and flexibly transferable modes. Meanwhile, existing emergency exits in buildings should connect to these priority routes providing targeted evacuation points.

6. Conclusion
The flooding situation of HCMC is an evidence of inappropriate urban expansion leading to increase in flooding vulnerability. Although climate change impacts are obvious, the rapid population growth and associated accommodation development are believed to be the key cause which has not been solved. This research found that the three new emerging districts (District 2, 9 and ThuDuc) are highly vulnerable to floods in the near future, but the local government still implements the plan for attracted investments in housing without a reference to the flood impacts evaluated by existing researches. This not only exposes more citizens and their assets to floods but also exacerbates the flooding situation of surrounding areas. Furthermore, the existing flood management relies on hard-engineered systems, and soft solutions, such as green-spaces for water absorption and reverting natural water ways and reservoirs integrated in parks for water storage have not been cooperated and deployed in practice. This also undermines urban resilience to extreme floods in the long term. Therefore, the city needs neater strategies for planning and management of spatial development to minimize the areas vulnerable to floods:

- creating more compact spaces in the central areas (Zone 1) protected by the current flooding management system; and
- offering more resilient spaces for new development areas (Zone 2), by improving resilience of transportation system.

Nevertheless, a similar combination of compact spaces and resilient spaces in emerging districts could also be incorporated into the existing developments, while sustainable drainage systems or underground water storage in buildings could also be included in the design to compensate for the former wetlands lost.

The findings in HCMC can be extended to the developing pattern of many coastal cities in Southeast Asia. For almost all these countries, economic development can be seen as a driving factor, expanding the cities despite the obvious disadvantages of topography and the negative interactions to the existing natural environment. Natural disasters, including floods, are likely to exceed present day predictions despite the improvement of forecasting technology. Therefore, these cities need to rethink how they can use the most naturally flood-resistant land (i.e. compact approach) to improve the overall resilience of new developments. As a contribution to urban planning and management in practice, this paper underlines the increasingly important role of urban planning and management to combat the future impacts of floods in coastal-mega cities. Furthermore, a combination of the compact city and resilient city, there is a potential topic for further research in planning and management of urban spatial development adaptive to flooding.

References

Angel, S., Sheppard, S.C. and Civco, D.L. (2005), *The Dynamics of Urban Expansion, Transport and Urban Development Department*, World Bank, Washington, DC.

Asia Development Bank – ADB (2010), *Ho Chi Minh City Adaptation to Climate Change*, ADB, Philippines, ISBN: 978-971-561-893-9.

Balica, S.F. and Wright, N.G. (2009), "A network of knowledge on applying an indicator-based methodology for minimizing flood vulnerability", *Hydrol Process*, Vol. 23 No. 20, pp. 2983-2986, doi: 10.1002/hyp.7424.

Balica, S.F., Wright, N.G. and Van deu Meulen, F. (2012), "A flood vulnerability index for coastal cities and its use in assessing climate change impacts", *Natural Hazards*, Vol. 64 No. 1, pp. 73-105, doi: 10.1007/s11069-012-0234-1.

Breheny, M. and Rookwood, R. (1993), "Planning the sustainable city region", in Blowers, A. (Ed.), *Planning for a Sustainable Environment, a Report by the Town and Country Planning Association*, Earthscan, London.

Burby, R.J., Deyle, R.E., Godschalk, D.R. and Olshansky, R.B. (2000), "Creating hazard resilient communities through land-use planning", *Natural Hazards Review*, Vol. 1 No. 2, pp. 99-106, doi: 10.1061/(ASCE)1527-6988 (2000)1:2(99).

Chapman, L. and Ryley, T. (2012), *Transport and Climate Change*, Emerald Group, London, Vol. 2, ISBN: 978-1-78052-440-5.

Cigler, B.A. (2007), *The "Big Questions" of Katrina and the 2005 Great Flood of New Orleans*, Pennsylvania State University, Harrisburg.

Commission of the European Community-CEC (1990), *Green Paper on the Urban Environment*, CEC, Brussels, EUR 12902 EN.

Dasgupta, S., Blankespoor, B. and Laplante, B. (2012), "Sea-level rise and coastal wetlands", Policy Research Working Paper (WPS6277), World Bank.

Desouza, K.C. and Flanery, T.H. (2013), "Designing, planning, and managing resilient cities: a conceptual framework", *Cities*, Vol. 35, pp. 89-99.Elsevier, doi: 10.1016/j.cities.2013.06.003.

Dutta, D. (2011), "An integrated tool for assessment of flood vulnerability of coastal cities to sea-level rise and potential socio- economic impacts: a case study in Bangkok – Thailand", *Hydrological Sciences Journal*, Vol. 56 No. 5, pp. 805-823, doi: 10.1080/02626667.2011.585611.

Elkin, T., McLaren, D. and Hilllman, M. (1991), *Reviving the City: Towards Sustainable Urban Development*, Friends of The Earth, London.

Fuchs, S., Kuhlicke, C. and Meyer, V. (2011), "Editorial for the special issue: vulnerability to natural hazards—the challenge of integration", *Natural Hazards*, Vol. 58 No. 2, pp. 609-619, doi: 10.1007/s11069-011-9825-5.

General Statistic Office of Vietnam – GSOV (2015), "Population of Ho Chi Minh from 1986-2015".

Hall, P. and Jones, M.T. (2011), *Urban and Regional Planning*, 5th ed., Rutledge, London, ISBN: 978-0-415-56654-4.

Haughton, G. and Hunter, C. (1994), *Sustainable Cities*, Jessica Kingsley Publishers, London.

Hong, T.D. (2011), "Flooding in Saigon", (translated from Vietnamese), available at: http://myweb.tiscali.co.uk/hbtdp/_private/Sai%20Gon%20ngap%20lut%20-%20Tran%20Dang%20Hong.htm

Hufschmidt, G. (2011), "A comparative analysis of several vulnerability concepts", *Natural Hazards*, Vol. 58 No. 2, pp. 621-643, doi: 10.1007/S11069-011-9823-7.

International Center for Environmental Management ICEM (2009), "Ho Chi Minh City adaptation to climate change", *Volume 2 – Main Report*, ICEM – ADB.

International Panel on Climate Change – IPCC (2014), "Climate change 2014: synthesis report", *Contribution of Working Groups I, II and III to the Fifth Assessment Report of the Intergovernmental Panel on Climate Change*, Core Writing Team, R.K. Pachauri and L.A. Meyer (Eds.), IPCC, Geneva, p. 151.

Jenks, M., Burton, E. and Williams, K. (1996), *The Compact City: A Sustainable Urban Form?*, E & FN Spon, London, ISBN 0419213007.

McLaren, D.P. (1992), "Compact or dispersed? Dilutions is no solution", in Willis, K.G., Turner, R.K. and Bateman, I.J (Eds), *Urban Planning and Management*, Edward Elgar Publishing, Cheltenham, pp. 43-59.

Matsumoto, T. (2012), "Compact city policies: a comparative assessment", OECD available at: www.oecd.org/greengrowth/compact-city-policies-9789264167865-en.htm

Ministry of Natural Resources and Environment of Vietnam – MNRE (2012), *Climate Change Scenarios for Vietnam*, Vietnam Publishing House of Natural Resources, Environment and Cartography, Hanoi.

Nguyen, H. (2015), "Ho Chi Minh City needs about VND 100,000 billion for flood defense (translated from Vietnamese)", VNEXpress, available at: http://vnexpress.net/tin-tuc/thoi-su/tp-hcm-can-100-000-ty-dong-de-chong-ngap-3281838.html (accessed 19 April 2017).

Nicholls, R.J. (2011), "A global ranking of port cities with high exposure to climate extremes", *Climatic Change*, Vol. 104, pp. 89-111, doi: 10.1007/s10584-010-9977-4.

Phi, H.L. (2013), "Urban flood in Ho Chi Minh City: causes and management strategy", *Vietnamese Journal of Construction Planning*, Vol. 63, pp. 26-29, Hanoi. (In Vietnamese).

Scheuer, S., Haase, D. and Meyer, V. (2010), "Exploring multi-criteria flood vulnerability by integrating economic, social and ecological dimensions of flood risk and coping capacity: from a starting point view towards an end point view of vulnerability", *Nat Hazards*, Vol. 58, pp. 731-751, doi: 10.1007/s11069-010-9666-7.

Smits, A.J.M., Nienhuis, P.H. and Saeijs, H.L.F. (2006), "Changing estuaries, changing views", *Hydrobiologia*, Vol. 565, pp. 339-355, doi: 10.1007/s10750-005-1924-4.

Southern Regional Hydro – Meteorological Center – SRHC (2015), "Water level and Precipitation Records 2010-2015", (data extracted).

Steering Center of Flood Control Program of Ho Chi Minh City – SCFC (2016), "Flooding reports for the events in 15th September 2015, and in 26th September 2016".

Storch, H. and Downes, N.K. (2011), "A scenario-based approach to assess Ho Chi Minh City urban development strategies against the impact of climate change", *Cities*, Vol. 28, pp. 517-526, doi: 10.1016/J.CITIES.2011.07.002.

Storch, H., Downes, N., Thinh, N.X., Thamm, H.P., Phi, H.L., Thuc, T., Thuan, N.T.H., Emberger, G., Goedecke, M., Welsch, M. and Schmid, M. (2009), "Adaptation planning framework to climate change for the urban area of Ho Chi Minh City, Vietnam", *Fifth Urban Research Symposium 2009*, World Bank, available at: http://siteresources.worldbank.org/INTURBANDEVELOPMENT/Resources/336387-1256566800920/6505269-1268260567624/Storch.pdf

TEDISouth (2010), "Transportation planning for Ho Chi Minh City by 2020".

Thinh, N.X., Bräuer, A. and Teucher, V. (2009), "Introduction into work package urban flooding of the BMBF megacity research project Ho Chi Minh City", *Environmental Informatics and Industrial Environmental Protection: Concepts, Methods and Tools*, Shaker Verlag, Berlin, ISBN: 978-3-8322-8397-1.

Thomas, L. and Cousin, W. (1996), "The compact city: a successful, desirable and achievable urban form?", Jenks, M., Burton, E. and Williams, K. (Eds), *The Compact City: A Sustainable Urban Form?*, E & FN Spon, London, pp. 53-65, ISBN 0419213007.

Tu, T.T. (2010), "Adaptation to flood risks in Ho Chi Minh City, Vietnam", *International Journal of Climate Change Strategies and Management*, Vol. 3 No. 1, pp. 61-67, doi: 10.1108/17568691111107943.

United Nations – UN (2012), "Population division", *World Urbanisation Prospects: The 2011 Revision*, Department of Economic and Social Affairs, United Nations, New York.

United Nations – UN (2014), *World Urbanization Prospects – Highlights*, UN, New York, NY.

Viet, L.V. (2008), "The urbanization and Climate Changes in Ho Chi Minh City", *Proceedings of the 10th Conference of Res. Inst. of Hydrology and Environment*, pp. 369-375.

Willroth, P., Revilla Diez, J. and Aruntai, N. (2010), "Modeling the economic vulnerability of households in the Phang-Nga province (Thailand) to natural disasters", *Nat Hazards*, Vol. 58, pp. 753-769, doi: 10.1007/s11069-010-9635-1.

World Bank - WB (2010), *Climate Risks and Adaptation in Asian Coastal Megacities: A Synthesis Report*, The World Bank, Washington, DC.

Zevenbergen, C. and Gersonius, B. (2007), "Challenges in urban flood management", in Ashley, R., Garvin, S., Pasche, E., Vassilopoulos, A. and Zevenbergen (Eds), *Advances in Urban Flood Management*, Taylor & Francis, New York, NY, pp. 1-11, doi: 10.1201/9780203945988.

Zevenbergen, C., Veerbeek, W., Gersonius, B. and VanHerk, S. (2008), "Challenges in UMF – travelling across spatial and temporal scales", *Flood Risk Management*, Vol. 1, pp. 81-88, Blackwell Publishing Ltd.

Corresponding author
Phan N. Duy can be contacted at: duyadm@gmail.com and duy.phannhut@uah.edu.vn

8

Displacement and climate change: improving planning policy and increasing community resilience

Tijana Crnčević
Institute of Architecture and Urban and Spatial Planning of Serbia,
Belgrade, Serbia, and

Violeta Orlović Lovren
Faculty of Philosophy, University of Belgrade, Belgrade, Serbia

Abstract

Purpose – The purpose of this paper is to present the major gaps in the field of planning policy and its implementation regarding climate change and disaster risk reduction (DRR), with special reference to the displacement of people, together with the knowledge needed to increase community resilience. The researched relations are illustrated by the example of Serbia. The Republic of Serbia has been faced with increasingly visible impacts of climate change in recent years – floods, heat waves, droughts and others. During the floods that hit Serbia in 2014, over 30,000 people experienced displacement. These events have triggered numerous efforts, both to repair the incurred damage and to analyze opportunities for prevention.

Design/methodology/approach – This research has used document analysis to investigate contemporary approaches defined by policies, programs and research reports regarding climate change and DRR, with special reference to the displacement of people. An analytical framework has been used to evaluate to what extent the planning policy framework in Serbia addresses these issues in the context of achieving resilient development. Secondary analysis of research data has been used to recognize the gaps and identify needs for increasing community resilience.

Findings – Based on the growing trends in projections of climate change as a result of induced natural disasters for the region in the future and international trends in coping with these issues, this paper argues that it is necessary to improve the implementation of the planning policy framework and the capacities of professionals and citizens, to reduce future displacement and increase community resilience to climate change. The key weaknesses found within DRR and the emergency management system in Serbia were the lack of an appropriate information base of the cadastre of risk zones and the lack of information and coordination of actors on the local to the national level. During the "pre-disaster" period, findings stress a weak partnership and capacity development practice at the local level, as well as between local responsible bodies and regional/national entities in charge of emergency management and DRR. The paper singles out the main preconditions for achieving effective resilient planning, so that such a plan can move "people away from marginal areas" and provide living conditions that are resilient.

Originality/value – This paper provides a comprehensive insight analysis of the relations between climate change and DRR, with special reference to the planning policy. Using the lessons learned from the recent climate-induced disaster with its implications on displacement, the paper identifies needs for strengthening capacities to establish more resilient communities in Serbia. The gaps and needs identified, as well as the recommendations provided, may be of value for neighboring countries as well, who face similar challenges in climate change adaptation and who need to increase disaster risk resilience.

Keywords Displacement, Climate change, Floods, Community resilience, Planning policy

1. Introduction

The issue of climate change today is an integral part of the contemporary global policy of sustainable development. Within the 17 Sustainable Development Goals (SDGs), Goal 13 directly promotes climate change issues, i.e. to "take urgent action to combat climate change and its impacts", while two others (Goals 9 and 11) indirectly address the issue by supporting resilient development (United Nations, 2015). To achieve these set goals, in December 2015, as a result of the 21st Conference of the Parties of the United Nations Framework Convention on Climate Change (UNFCCC), the Paris Agreement was adopted, after which all signatories undertook the obligation to limit the increase in global temperatures up to 1.5 degrees (194 countries signed, including the Republic of Serbia, and 141 ratified the agreement) (http://unfccc.int/paris_agreement/items/9485.php).

The aim of establishing resilient human settlements is the protection of the population and the prevention of displacement and migration as a result of the impacts of climate change. The Intergovernmental Panel on Climate Change (IPCC) in its report from 1990 indicates that one of the main impacts of climate change may be on human migration (WMO, United Nations Environment Programme (UNEP), IPPC, 1990). The focus of this latest report is on the adaptive or vulnerability capacities of populations to climate change rather than on migration. The report indicates that climate change may result in the "displacement of people" and that "populations that lack the resources for planned migration experience higher exposure to extreme weather events, particularly in developing countries with low income" (IPCC, 2014a, p. 16). However, as pointed out, "there are no empirical data on the impact of climate change on the population" (Brown, 2008, p. 36), which stresses the fact that migration is the result of inadequate planning and, when migration continues following a natural disaster, it is "a reflection of the state's deficient response, rather than the natural hazard impact" (Raleigh and Jordan, 2010, p. 112). Thus far, the Internal Displacement Monitoring Center (IDMC) reports that permanent environmental migration has already occurred – in 2014, more than 19 million people from 100 countries were displaced (IDMC, 2016). Furthermore, predictions also indicate that the number of people affected by floods will be between 10 and 25 million by the year 2050 and between 40 and 140 million by 2100, depending on the future scenarios consulted (Nicholls and Lowe, 2004).

Even in fields of growing theoretical, policy and practical interest, such as disaster resilience and environmental displacement and migration, issues related to the impacts of climate change on humans are often analyzed from a one-sided perspective – as a consequence of predominantly environmental factors or as an effect of destructive human behavior toward the environment (emission of gases, deforestation, etc.). In addition to the largely recognized difficulty of viewing these phenomena within a holistic framework, there is also the rightly emphasized need to pay particular attention to the diversity of contexts in societies and communities around the globe. Overall, whether it comes to displacement and migration as a result of climate change or environmental impact, contemporary strategies are considering the distress migration patterns or migration as a result of natural disasters.

These migration models carry two basic assumptions: the first is that the migration is internal rather than international and that such migration is temporary (Raleigh *et al.*, 2008). Further, there is no legislative response "either literally or figuratively" (Brown, 2008, p. 36) except that Sweden is including within its immigration policy, environmental migrants as "persons in need of protection" (Brown, 2008). What should be emphasized is that displacement is mostly forced, while migration is mainly voluntary (Yonetani, 2017), and thus, within the Sendai Framework for Disaster Risk Reduction (SFDRR), displacement is also indicated as "major human consequences and drives of disaster risk" (Yonetani, 2017). Therefore, building capacities for disaster risk reduction (DRR) aims at reducing the need for and potential consequences of displacement. While DRR establishes a set of well-planned, long-term strategic measures and activities, displacement comes as an *ad hoc*, reactive measure, taken when an emergency has already occurred.

A modern approach to DRR requires human responsibility and building awareness of human influence and its contributions both to vulnerability and to resilience to disasters. "A natural hazard" (such as an approaching storm) only becomes a "natural disaster" if a community is particularly *vulnerable* to its impact. A community's vulnerability, then, is a function of its *exposure* to climatic conditions (such as a coastal location) and the community's *adaptive capacity* (the capacity of a particular community to weather the worst of the storm and recover after it) (Brown, 2008, p. 18). A paradigm shift is required in the approach to the individual and the community role in coping with disasters – from the concept of "passive receivers of help" and "victims", toward active, knowledgeable, adaptive citizens.

The growing discussion of the concept of the Anthropocene, which places humans as a "global geological force in their own right" (Steffen *et al.*, 2011), sheds light on human power, rather than on human vulnerability. Accordingly, it is acknowledged that human activity brings about huge geological (and consequently climate) change, using – or abusing – knowledge and technology. Despite different paradigms behind the pictures of vulnerable and powerful humankind, both call for reflection on our role and responsibility toward the physical and social environment of the present and the future, requiring "wise technology and knowledge" (www.ecomodernism.org).

Living in a "risk society" (Beck, 2001), we face a reality in which the distribution of risks, as well as capacities to cope, is not equal in the modern world (Orlovic-Lovren and Pejatovic, 2015). In terms of environmental/climate displacement and migrations, it is necessary to look at the wider context of sociodemographic processes, such as migrations from rural to urban areas, accelerating in particular in low- and middle-income countries, as well as the "brain-drain" – both characteristic mostly of developing countries. (Brown, 2008). One of the recent global trends is the increasing number of migrants being forced to leave their homes under the risks of armed conflicts. Unlike those forced by natural disasters, these migrants tend to move externally, to faraway destinations – usually to more developed countries. These "altered patterns of forced migrations" (Miladinovic, 2016) bring new requirements in terms of adaptation to the new environment and culture, as well as relationships with the new community and the development of social networks that can support the integration process.

Adaptation to climate risks is highly influenced by the ability of the system – collective or individual – to learn. Learning to understand the nature, causes and consequences of risks, as well as the – individual or collective – needs and opportunities to prevent or reduce its influence, are among the key segments of building resilience, a part of "the ability of individuals, communities, organizations or countries exposed to disasters and crises and underlying vulnerabilities to anticipate, reduce the impact of, cope with, and recover from the effects of shocks and stresses without compromising their long term prospects"

(IFRC, 2015). The new paradigm that promotes the Sendai Framework for Disaster Risk Reduction 2015-2030 (UNISDR, 2015) presents DRR and resilience as the key elements for achieving sustainable human development with focuses toward creating "safer living environments and for improving building codes [. . .]" with continuous strengthening of local communities (IFRC, 2016, p. 22). The message from all relevant global strategies – related to climate change, disaster risk reduction or sustainable development – is strong and clear in advocating for improving capacities for adapting to and coping with climate change, risks and disasters.

Taking into account the global policies and current trends within climate change and DRR with respect to the displacement of people, the purpose of this paper is to present the status of the Republic of Serbia and its capacities regarding resilient development, with special reference to the planning policy and capacities of professionals and citizens.

2. Methodology

Starting from the aim of this paper – to present the major gaps in the field of planning policy and its implementation regarding climate change and DRR in Serbia, with special reference to the displacement of people – a document analysis has been performed. Global policy documents have been analyzed to identify contemporary approaches defined by strategies, programs and research reports in the context of achieving resilient sustainable development. The analytical framework is then used to evaluate to what extent the planning policy framework in Serbia addresses issues of climate change resilience and promotes measures toward strengthening community resilience to climate change.

In addition, a secondary analysis of research data has been used to recognize the gaps and identify needs for increasing community resilience. The researched relations are illustrated through the example of Serbia.

3. Serbia: case study

3.1 The context

The changes in climate conditions within the past decades in Serbia are evident. There was an increase in episodes of stronger rainfalls, and even though the changes in overall quantities were small, there was a warming trend in terms of the increase of extreme high temperatures and extended periods of warm weather and droughts (which especially endangered North Vojvodina) and increased periods of very low temperatures (Official Gazette of the Republic of Serbia, 2011; United Nations Development Programme (UNDP), 2016). In the context of expected climate changes of climatic conditions, taking into account different scenarios, the report stresses the possibility of an increase in temperature, with more pronounced warming during summer and autumn that would exceed 4°C by the end of the century. In relation to precipitation, a positive trend is expected until the year 2040, but which by the end of the century will reduce to negative values, with the largest deficit in summer (Republika Srbija (RE), 2017). All scenarios of climate change indicate that Serbia and the Western Balkans face a high probability of rising temperatures in the future, with frequent and prolonged drought and wildfires expected. These changes will be accompanied by prolonged and intense rainfall and melting snow, which will result in the emergence of a large pour point, landslides and flood sediment (IPCC, 2014b; UNDP, 2016).

Vulnerable areas in Serbia cover 57.33 per cent or 50656.87 km^2, of which over 20 per cent are areas vulnerable to drought, followed by seismic hazards (18.55 per cent VIII-IX MCS), potential floodable areas (17.20 per cent), landslide hazard areas (15.08 per cent), forest fires (3.57 per cent), erosion (3.76 per cent) and seismic hazards of IX-X MCS (1.26 per cent) (Dragicevic et al., 2011). Compared to other countries within the region, the population of

Serbia is most at risk of flooding – the potential flood areas in the Republic of Serbia cover about 1.6 million hectares, with about 500 larger settlements and 515 industrial facilities, 680 km of railways and 4,000 km of roads within 30% of the country is agricultural land (UNDP, 2016; Official Gazette of the Republic of Serbia, 2010). At the end of the twentieth century, the frequency and intensity of floods in Serbia increased. From 1980 to1990, 2,000 natural disasters were recorded, while in the last decade of the twentieth century, the figure rose to 2,800 (Official Gazette of the Republic of Serbia, 2011). In 2014 and 2015, 2.8 million people were affected by floods that hit the region, with damages reaching $4.63bn (UNDP, 2016). In May 2014 alone, floods hit 38 towns and municipalities in central and western Serbia – around 20 per cent of the population of Serbia and over 30,000 people were displaced. The total damage from the floods was mainly concentrated on manufacturing activities (€1,070m/70 per cent of total damage), then on social services (€242m/16 per cent) and infrastructure (€192m/12 per cent) (UN, EC, WBG, 2014). This disaster caused a recession in the economy of Serbia, which in 2014 recorded a decline of 1.8 per cent rather than the expected growth of 0.5 per cent (Bijelić and Lazarević, 2015). In the same year, four months later, floods hit the Bor region in eastern Serbia, causing damage in the municipalities of Negotin, Kladovo, Majdanpek, Tekija, Grabovica, Brza Palanka and Boljetin, so that total damage from flooding in 2014 was estimated at 4.8 per cent of the GDP of Serbia (UN, EC, WBG, 2014).

Because climate change and displacement are both global and context-specific phenomena, defined by "[...] various interacting factors such as social, environmental, political, climate, cultural, developmental and physical aspects" (Upadhyay et al., 2015), it is necessary to put our analysis in the context of the characteristics and recent developments in the Serbian society.

The political, social and economic turbulences that have taken place in Serbia since the 1990s, including the influences of armed conflicts on the territory of former Yugoslavia and the bombing of the Serbian territory in 1999, certainly contribute significant impacts to the vulnerability of its citizens. Keeping in mind the consequences of international economic sanctions by the international community and hyperinflation, as well as the duration of the transition (from the socialist political and economic regime), this period is rightly marked as "a prolonged period of crisis" in Serbia (Orlovic-Lovren and Pejatovic, 2015).

In addition to this general trend, discrepancies are also visible within the country – from the data on poverty and demographic characteristics of rural and urban areas, and in particular between the capital (Belgrade) and the rest of the country. The overall increase in poverty in the Republic of Serbia was generated as a result of increased poverty in rural areas. The poor are the least represented in northern parts of the country, including Belgrade. A huge disproportion may be found in terms of the demographic volume and characteristics between Belgrade and other cities and regions in Serbia, affecting capacities for sustainable development and the resilience of its inhabitants (Official Gazette of the Republic of Serbia, 2012) (Table I).

Table I. Percentage of poverty based on type of settlements –absolute poverty line

Type of settlement	2006	2007	2008	2009	2010
Serbia	8.8	8.3	6.1	6.9	9.2
Urban areas	5.3	6.0	5.0	4.9	5.7
Other areas	13.3	11.2	7.5	9.6	13.6

Source: Statistical office of Serbia: www.stat.gov.rs

As is now largely evident in different regions of the world, along with poverty and overpopulation, there is an increase in exposure to disaster risks, which then further affects the vulnerability of people. Living in areas close to river banks and in poorly constructed houses, the poor are most often affected by floods and earthquakes. Not only improvements in infrastructure but also in environmental and planning legislation and procedures are strongly needed, especially in those cases where the situation is further complicated by the lack of the capacity of these people to participate and to actively contribute to sustainable solutions. In this context, the goal of contemporary planning is not only to "produce a perfect document but to develop a process" (ADPC, 2010, p. 3) because "what really counts is how the plan is prepared" (ADPC, 2010).

One of the demographic trends typical of Serbia since the last decade of the twentieth century is the waves of refugees and internally displaced people, caused mainly by the conflicts in the former Yugoslavia. Data show that at the end of the past decade, there were 86,336 refugees and 205,835 displaced people in Serbia (Krstic *et al.*, 2010). In addition to their unfavorable social status, they often belong to groups of poor and socially excluded people who live in inadequate settlements or parts of settlements. The situation is even worse in the case of the Roma and persons with disabilities because they are particularly affected by poverty, as a result of the accumulation of various risks. Of all the social and economic risks, the dominant one for these groups is unemployment.

In terms of aging and depopulation, rural and mountainous areas are particularly vulnerable in Serbia. Entire regions within Serbia have been affected by general economic trends, as well as by additional demographic trends, first of all by the migration of the younger and more educated toward urban centers. These regions outline the unfortunate "exclusion curves", such as the area of East Serbia (Krstic *et al.*, 2010). The vulnerability of the population in less developed areas and those belonging to poverty circles and marginalized groups contributes to their overall vulnerability to disaster risks in the country – on individual, family and community levels.

When examining general migration trends, it is important to mention those related to the recent waves of migrants from conflict zones or those searching for better economic conditions. According to data from the beginning of 2017, there were around 7,700 asylum-seekers and migrants in Serbia, mainly from Afghanistan, Iraq and Syria, who were sheltered (around 85 per cent of them) in 17 facilities provided by the government (UNHCR Serbia update, 2017). Most of them are children (47 per cent) and for some of them (age 7-15) in some of the reception centers, there are organized education programs (UNHCR Serbia update, 2017). Though the majority of these migrants consider Serbia only as a transit route and do not intend to stay here for a long time, this problem may be seen as indirectly related – but still relevant – for the mapping of the characteristics of vulnerability to disaster risks and the need to increase the resilience of the population in Serbia.

3.2 Policy framework

Taking into consideration that the main orientation of the Republic of Serbia is harmonization within the European Union (EU) policy, Serbia in the past decade has been undergoing changes to improve the legal and planning framework in all segments. Since March 2012, after receiving the status of candidate for EU membership, six chapters of the acquis have been opened. Chapter 27, which covers the field of environmental protection, is being considering under the project "Further implementation of the National Strategy for approximation in the field of environmental protection" that should be completed by October 2018. Regarding the context of climate change, special activities are directed primarily in the domain of the legal framework and the harmonization of development policies. It should be noted that Serbia has signed all relevant frameworks regarding the climate change issue (Kyoto Protocol, UNFCCC, Hyogo Framework for Action, Sendai Framework for Disaster Risk Reduction 2015-2030). An overview of national frameworks of relevance to climate change, indicating the main scope regarding the selected strategies with special reference to DRR, is provided in Table II.

Regarding DRR, the following programs were adopted: *National Program to Manage the Risk of Natural Disasters* (2014) and the *Action Plan for the Implementation of the National Program for Risk Management of Natural Disasters* (2016). The adoption of the following law is in progress: *Law on Reducing the Risks of Natural and Other Disasters and Emergency Management*. These documents provide solid foundations for integrating DRR principles in the policies and practices of development at all levels – from local to national. Local governments are seen as the main contributors to DRR and are expected to actively participate in the development of plans for DRR based on previously completed risk assessments.

These strategic frameworks suggest that the Republic of Serbia's access to the issue of climate change is comprehensive, both in the fields of mitigation and adaptation. Signing the global frameworks that address the climate change issue and establishing a strategic DRR framework are the first steps in the implementation of commitments. In accordance with the strategic documents described above, the legal basis provides support regarding the issue of climate change. The bases for the inclusion of climate change issues in spatial and urban planning are given in the *Law on the Spatial Plan of the Republic of Serbia 2010-2014-2021* (Official Gazette of the RS, no. 88/10), the *Law on Strategic Environmental Assessment* (Official Gazette of the RS, no. 88/2010) and the *Law on Environmental Impact Assessment* (Official Gazette of the RS, no. 135/2004.36/ 2009), providing the scope for the inclusion of the issue within planning and assessment. Indirectly, the climate change issue is represented in terms of promoting measures for adaptation and mitigation, such as: development of areas within the program NATURA 2000, the development of ecological networks after the Law on Nature Protection (Official gazette. RS no. 36/2009, 88/2010), through the rational use of non-renewable natural resources and renewable energy sources and by promoting energy efficiency under the Law on Planning and Construction (Official gazette of RS No. 79/09, 81/09-correction, 64/10-US, 24/11, 121/12, 42/13-US, 50/13-US, 98/13-US, 132/14 and 145/14). In addition, the Law on Forests (Official Gazette of the RS, no. 30/10 i 93/12) acknowledges the favorable impact of forests on the climate and protection measures in the case of disruption of the biological balance and also the serious damage within forest ecosystems caused by natural disasters, while within the Law on Environmental Protection (Official Gazette of the RS, no. 135/2004), the climate issues have not been considered directly. Together with a strategic framework, these legal bases provide the main preconditions for the inclusion of climate change issues within planning. However, it is stressed that the "[. . .] climate change problem area has not been systemically treated in the SEA, i.e. plans" and therefore "the current legal framework [. . .] is not fully supportive of this theme" (Crncevic *et al.*, 2011, p. 19). A main limitation is that the current legal and planning bases are "missing an adequate response to the potential risks of climate change, especially in the fields of impact research and implementation of the planning instrument" (Crncevic *et al.*, 2016, p. 168). However, as a main potential, it is noted that "plans give full support to the measures such as increasing protected areas, developing ecological networks, ecosystem protection, monitoring of invasive species, and planning measures for their

Table II. Review of strategic framework of the importance of climate change issues

National Strategy for Sustainable Development and *Action Plan for the period 2009-2017* (Official Gazette of the RS, 57/2008)	Supports sectoral actions for reducing GHG emissions and development of a plan of adaptation within the economic sector; contains recommendations for the development and implementation of the program of risk assessment that would include risk maps, development of a database of landslides and evaluation of risk mitigation measures; within a separate chapter "Natural disasters – floods, landslides, fires, earthquakes" promotes flood protection and institutionalization of civil protection
National Program for Environmental Protection for the period 2010-2019. (Official Gazette of the RS, 12/2010)	Establishes a framework for action on climate change/ defines the actions needed to mitigate the effects of climate change
National Strategy for Inclusion of the Republic of Serbia in the Clean Development Mechanism (CDM) (Official Gazette of the RS, 8/2010)	Identifies priority modes and options for inclusion in the Clean Development Mechanism for the sectors of agriculture, forestry and waste management
Strategy for Forest Development of RS (Official Gazette of the RS, 59/2006)	Defines priorities, including stopping the further reduction of forested areas, and emphasizes the role that the forestry sector has in the sustainable development of the Republic of Serbia, as well as the importance of forests for flood prevention
Strategy of Biodiversity for the period 2011-2018 (RS, 2011)	Defines strategic areas, objectives and activities for biodiversity protection, as well as an Action plan. Supports the increase of capacity of relevant institutions to monitor and forecast the impact of climate change on biodiversity, capacity building in all sectors and in public and the development of a national strategy and mechanisms for better understanding of planning, regarding the impacts of climate change
National Strategy for Disaster Risk Reduction and Protection and Rescue in Emergency Situations (Official Gazette of the RS, 86/2011)	The national strategy ensures fulfillment of the recommendations of the European Union for the development of a national system of protection: the establishment of institutional, organizational and personal conditions for the implementation of protection in emergency situations
National Strategy for Sustainable Use of Natural Resources and Properties (Official Gazette of the RS, 33/2012)	Establishes priorities among the obligations for developing a system for monitoring the effects of climate change on biodiversity and ecosystems: increasing the territory under protection, the establishment of a national ecological network and the preparations for adopting the protection program Natura 2000; promoting adaptive management and the introduction of principles of adaptive planning in the management plans of protected areas, biodiversity protection and sustainable use of biological resources
Energy Development Strategy until 2025 (Official Gazette of the RS, 101/2015)	Defines the conditions for the promotion of energy efficiency within activities related to energy and energy consumption; supports the use of renewable energy and promotes environmental protection
Strategy for Agriculture and Rural Development of the Republic of Serbia for the period 2014-2024 (Official Gazette of the RS, 85/2014)	Respecting the issue of climate change, establishes the basis for a new policy in the field of agriculture, in line with the European framework and support for agriculture

suppression, promotion of biodiversity values and promotion of the development of the monitoring system" (Crncevic, 2013, p. 80).

The national policy for protection from natural disasters is defined in the *Law on Emergency Situations* (Official Gazette of the RS, no. 111/09, 92/11, 93/12) and includes the requirement to establish protection conditions for the prevention of natural disasters (floods, droughts, torrents, storms, heavy rain, atmospheric discharges, hail or landslides, avalanches and snow layers, extreme air temperatures, ice accumulation on the watercourse) and their inclusion in planning documents. The Law introduces the requirements for risk assessment, which is prescribed in greater detail within a further set of laws and regulations[1]. Together with the *National Strategy for Protection and Rescue in Emergency Situations* (Official Gazette of the RS, no. 86/2011), this represents the main framework for implementing the Hyogo framework. The *Law on Emergency Situations* provides a solid base for an integrated approach regarding emergency management, but does not deal much with community resilience. One of the main weaknesses is that it puts greater emphasis on response than on prevention. This has been improved by the new *Law on Reducing the Risks from Natural and Other Disaster and Emergency Management (*in progress), together with the inclusion of the Sendai principles. The *Law on Planning and Construction* (Official Gazette of the RS, no. 79/09, 81/09-correction, 64/10, 24/11-US, 121/12, 42/13-US, 50/13-US, 98/13-US,132/14 and 145/14) after *Regulations of the Content, Method and Procedures of Preparation of the Planning Documents* (Official Gazette of the RS, no. 31/10, 69/10 and 16/11) establishes the obligation within planning documents to display zones of vulnerability and activities on environmental protection, natural and cultural heritage and to organize the space of interest for national defense and protection against natural disasters within planning documents. The Water Act (Official Gazette of the RS, no. 30/2010, 93/2012, 101/2016), which defines the obligation to produce a general and an operational plan for flood protection, has been revised regarding the obligation for the production of erosion zones, for which the relevant ministry is now responsible, and not local authorities, as it was previously defined. It should also be emphasized that the legislative framework includes the following, in the planning of the DRR:

- Law on Meteorological and Hydrological Activities (Official Gazette of the RS, no. 88/2010) – regulates the hydrological and meteorological activities, including activities related to hydrological disasters and making weather forecasts for early warning and provides information for climate and weather risk assessment;
- Law on Fire Protection (Official Gazette of the RS, no. 111/2009, 20/2015) establishes the role of local governments in the prevention and formulation of plans for fire protection and establishes the obligations of all participants in fire protection and prevention of risks;
- Law on Local Government Act (Official Gazette of the RS, no. 129/2007, 83/2014) directly promotes reducing the risk from disaster in the sense that municipalities "[. . .] shall, through its units, in accordance with the Constitution and relevant laws, to organize protection from disasters and fire protection and create conditions for their prevention and mitigation"; further, the law (Article 20) also establishes the responsibility of local government to identify areas threatened by erosion; and
- Law on Insurance (Official Gazette of the RS, no. 139/2014) defines property insurance against fire and natural disasters.

In regard to public participation within the planning process, it should be stressed that the participation is limited to public hearings (Law on Planning and Construction, articles 45a, 50, 51) and that the Law does "not oblige the developer of the Plan to cooperate with the local community and civil society" (Petovar and Jokic, 2011, p. 10). On the other hand, relevant authorities (organizations and public companies authorized to determine special conditions for the protection and regulation and construction of buildings, for protection of water courses, hydrometeorological and seismological institutions, etc.) are present throughout the entire process. As illustrated in the next chapter, the lack of participation of citizens is not only related to the lack of existing legal mechanisms but also to the unavailability of information, as well as to the lack of programs which would systematically develop their capacities to participate.

A direct response to the floods in May 2014 was the adoption of the *Law on Elimination of Consequences of Floods in the Republic of Serbia* (Official Gazette of RS, no. 75/14, 64/15), which aims to regulate the elimination of the consequences of floods, landslides and activate the areas affected by the floods. In 2016, to secure the implementation of risk reduction, the Office for the Management of Public Investments was established. The Office is tasked to coordinate actions to reduce risks and is in accordance with the Sendai Framework. Further, in 2016, in cooperation with and with financial assistance from the EU, the National Strategy to Combat Climate Change with an action plan was prepared throughout the period 2020-2050, as well as the assessment of emission reductions by 2070. In addition, work is being done toward creating an innovation planning base for the areas hit by floods. Currently, the procedure includes the "Spatial plan of exploitation of the Kolubara lignite basin, DRAFT" and risk assessment for Obrenovac, the *Assessment of Vulnerability to Natural Disasters and Other Accidents in the Obrenovac Municipality* (RS, City of Belgrade, Municipality Obrenovac, 2015), as well as a project directed toward defining needs and gaps within the current framework.

3.3 Gaps and needs to increase resilience to disasters in Serbia
The dramatic effects of floods and landslides that hit Serbia and the region in 2014 contributed to the increased attention of international donors, regional and national decision-makers, as well as of the research community. A number of analyses have been supported by international organizations. Within the Environment and Security (ENVSEC) Initiative, which is a partnership of the Organization for Security and Co-operation in Europe (OSCE), UNDP, UNEP, United Nations Economic Commission for Europe (UNECE) and Regional Environmental Center for Central and Eastern Europe (REC), and with The North Atlantic Treaty Organization (NATO) as an associate partner, there were a number of studies and projects aimed at contributing to reduced environment and security risks in countries in Central Asia, Eastern Europe, Southeast Europe and the South Caucasus.

Under the auspices of this initiative, during the summer of 2014, an assessment of needs for public education for DRR was performed at the local level in Serbia. Besides a survey with experts from competent institutions and organizations, the assessment comprised interviews with representatives of the local government, headquarters for emergency management and nongovernmental organizations from four municipalities that had experienced floods and landslides in previous months. Qualitative analysis of the data shows that the key obstacles to effective emergency response as perceived by these communities were the lack of communication and a timely approach and access for all citizens, a lack of efficient coordination between local and national units in charge of emergency management, the lack of technical equipment and a lack of trained professionals and volunteers for rescue and response (Orlovic-Lovren, 2014).

Research findings from a study performed with the support of OSCE, on gender analysis of the impact of the 2014 floods in Serbia, offer quite a similar picture of existing weaknesses. According to data collected through focus group discussions and interviews, community members (from four affected areas in Serbia) were not prepared for the emergency owing to the lack of information on the "course of action that should be taken in the event of emergency [...]. they had no access to information about what was going on [...] and what they could expect"; only 9 out of 78 respondents report that they were informed about the threat and evacuation, and all of them obtained that information from informal networks (Bacanovic, 2014).

According to the findings of other studies, the lack of quality information on disaster risks and its timely and wide dissemination is obviously common to the majority of municipalities in Serbia. Weaknesses in sharing information on the local risk profile are related to the fact that there is a lack of local risk assessment, and again, to the fact that the national risk assessment has still not been completed for Serbia. As pointed out, "the current level of integration of climate change into sectoral and overall development strategy, the level of knowledge, institutional and individual capacities, available technology and financial resources at the national level and involvement of local governments, despite numerous activities and efforts are not sufficient for an effective and rapid response to the problem of climate change" [Republika Srbija (RE), 2017].

In regards to the planning system of DRR, the limitations are owing to the requirements of the Law on Water, under which flood zones are to be determined by the state through its public water companies, but that work is still in its beginning stages, and the question is when it will be completed [Republika Srbija (RE), 2017]. The requirement to develop risk maps, following the Law on Emergency Situations, which includes the risk of flooding, is impossible to fulfill without a predetermined flood zone [Republika Srbija (RE), 2017]. Weaknesses related to the planning policy are owing to the lack of conditions for the consistent application of regulations, as well as the lack of a comprehensive database on the spatial distribution of certain natural disasters and accidents, and the determination of potentially critical zones (landslide, torrential streams, pollutants) – cadastre of risk zones, inadequate organization and implementation of preventive measures, the unavailability of specialized cadastre and the absence of a comprehensive risk map (RS, 2010, Stefanovic and Gavrilovic, 2014).

Evidence from the recent analyses on disaster risk resilience in municipalities in Serbia clearly illustrates the relation between the lack of plans and other prevention measures and the vulnerability of communities: while 11 out of 13 municipalities that are included in the research have completed plans for protection from floods, two other municipalities where those plans are missing belong to the group with a high vulnerability to floods (Andjelkovic and Kovac, 2016). The same study indicates once again that coordination between the local headquarters with the national bodies in charge of emergency management and DRR is not satisfactory; while they rely strongly on support from the national level in case of emergencies, local resources, and in particular citizens who want to help and civil protection trustees who should be actively engaged, are not adequately trained for that in most communities (Andjelkovic and Kovac, 2016).

Weaknesses related to information and the capacities of DRR actors and citizens in prevention are naturally reflected in the emergency response. Placing our focus on the temporary displacement during the last huge disaster induced by climate factors in Serbia, we may find that the process of evacuation of citizens from their homes in flooded areas, according to available evidence, was also followed by the lack of accurate information, which increased the uncertainty and anxiety of temporarily displaced people. Their participation in the information and decision-making flow was also characterized by traditional gender patterns: both during the evacuation as well as during their life in collective centers, men were typically more active and more informed than women, with the tendency to proactively seek information and the opportunity to go home and start working on cleaning and reconstruction (Bacanovic, 2014).

Uneven dissemination was characteristic not only for the process of information but also for the psychological support to displaced people: as noted, support was mainly provided for residents of collective centers in the capital city and not in other flooded areas (Bacanovic, 2014). In addition, the information and other forms of support provided in centers – mostly by volunteers – were not available equally to all, owing to illiteracy and language barriers, present first of all in ethnic minority groups, such as the Roma (Bacanovic, 2014).

As shown by this and other studies performed after the floods in 2014, volunteers and other participants from the civic sector play a significant role in emergency response; besides the activists from the Red Cross, the Mountain Rescue organization and civil protection trustees, who usually cooperate with the local headquarters for emergency management, other volunteers were perceived by respondents as valuable support, but groups which help would have been much more effective if they had been trained (Orlovic-Lovren, 2014; Andjelkovic and Kovac, 2016). While their activity is well recognized and intensive in times of response to disaster, they are much less engaged and involved in the prevention phases. In the response phase, there is also, as evidenced by studies, the strong influence of social capital to the coping and recovery capacities of communities; strong ties and trust in family, neighbors and friends seem to be one of the specific characteristics of importance for disaster resilience in Serbia (Andjelkovic and Kovac, 2016).

4. Conclusions
As shown by the data and tendencies evidenced by various studies, some of the key weaknesses of the DRR and emergency management system in Serbia are the lack of information and coordination of actors from the local to the national level. As protection against natural disasters must be based on reliable data, the main weakness within planning is found to be that it is missing an appropriate information base of the cadastre of risk zones. While during emergency response there is an important role of non-state (civic or private sector) actors, in the "pre-disaster" period, there is a weak partnership and capacity development practice at the local level, as well as between the locally responsible bodies and the regional/national entities in charge of emergency management and DRR.

Lack of access to important information and to engagement in resilience development activities is particularly visible among socially vulnerable groups, such as the elderly, persons with disabilities and illiterate or semi-literate citizens. Traditional gender patterns contribute to the unequal status of women, in particular of single mothers and those with lower education and who are unemployed, in both prevention and response to disasters.

As revealed by several studies, the key factors contributing to the vulnerability of the population in Serbia to climate-induced disasters are a lack of a systematic approach to information and of participation of citizens in programs aimed at developing the capacities of all actors, and in particular in the training of volunteers and civic organizations.

Taking into account the gaps identified by the recent studies in Serbia, reviewed above, efforts need to be undertaken – or continued – to improve resilience to climate-induced disasters. Such efforts should include the following:

- *permanent outreach campaigns aimed at raising awareness of people* using global and national policy instruments and documents which are carefully "translated" into their "language" – taking into account their understanding and specific needs so that the communication might stimulate their awareness and activity; these efforts should include outreach in risk awareness and DRR for refugees from foreign countries and for minority communities, such as the Roma;

- *participatory data collection and planning* – systematic information about the public, based on national and local risk maps and disaster reduction and emergency response plans; using local sources and knowledge and implementing a gender-differentiated approach in collecting and disseminating information; involving citizens in the planning and decision-making process in this field;

- *continuously tailored training/non-formal learning activities* – training of citizens at the community level in DRR and climate change adaptation/mitigation measures; training of civil protection trustees and members of the local emergency management teams, as well as of volunteers; training and partnerships with journalists and the media as an important precondition for developing community resilience in all its aspects;

- *mainstreaming DRR in education curricula and institutional development at all levels* – both students and teachers as well as the staff of educational institutions need to be included in education and practical activities/campaigns within the education system and in cooperation with communities; and

- *strengthening of partnerships* between research, experts and education institutions with DRR and emergency management units as well as with the nongovernmental and private sector in developing resilience at the local and national level.

Recognizing DRR methods and techniques as a part of the adaptive strategy in the planning process will be a step forward in achieving resilient development and reducing displacement. As has been stated, "moving away from the impacts of a disaster – or projected future disasters – is a rational and long-standing adaptation strategy" (IFRC, 2016, p. 22) that can be seen as a framework within future planning and the building of resilient living environments in Serbia. Achieving a "long-standing adaptation strategy" requires the continual improvement of the national planning policy framework that in Serbia, as has been stressed, is in a process of revision, with the EU and global framework presenting a valuable base. However, within related regulations covering climate change and DRR, the response is slightly slower and still needs coordination and harmonization, not only within the national framework (i.e. vertical coordination), but also within the legal framework (horizontal adjustments). The main precondition regarding the planning policy framework in Serbia for achieving resilient development and reducing displacement is a reliable information base that empowers the focus of the entire planning process and enables the implementation of all global, regional, national and local frameworks related to climate change and DRR, together with informed and involved professionals and citizens. In order to develop and build up planning as a process, strengthening monitoring towards continual improvement and innovation of the information base, standards, institutional and, as well, the capacities of professionals and citizens within the planning process will be a crucial task. Effective monitoring will provide not only an effective response and support to process development but also a permanent measure needed to protect vulnerable and risk zones from development and "move people away from marginal areas" to provide living conditions that are resilient. Finalizing the national risk assessment (expected next year) as a platform for developing local risk assessments and participatory planning processes at all the levels of governance, as well as further international support to national and local decision-makers and stakeholders, would be of great value for developing and modifying models of public participation which would lead toward strengthening resilience and adaptation to climate change in accordance with local needs and characteristics.

Note

1. Regulation of the content and manner of preparation of plans for protection and rescue in emergency case situations, Official Gazette of the RS, no. 8/11, Guidance on the methodology for the development of risk assessment and plan of protection and rescue in emergency situations, Official Gazette of the RS, no. 96/12, Rules on the method of preparation and content of contingency plans, Official Gazette of the RS, no. 82/12, Regulation of mandatory means and equipment for personal, mutual and collective protection from adverse weather conditions and other disasters, Official Gazette of the RS, no. 3/11, Rules on the organization and methods of use of socialized civil protection units, Official Gazette of the RS, no. 26/11, Regulation of the engagement of staff for the purposes of protection and rescue mode for realization of rights for compensation for the use of the same, Official Gazette of the RS, no. 10/13, 92/11, 93/12.

References

Andjelkovic, B. and Kovac, M. (2016), "Social capital: the invisible face of resilience", *Human Development Report Serbia*, UNDP Serbia, Belgrade.

Asian Disaster Preparedness Center (ADPC) (2010), "Urban governance and community resilience guides", *Compiled by Christine Apikul Technical Team NMSI Arambepola; Anisur Rahman; Padma Karunaratne; Anggraini Dewi; Khondoker Golam Tawhid; Gabrielle Iglesias Contributors Ernesto Elcamel; Muhammad Murad Billah; Kittiphong Phongsapan; Rendy Dwi Katiko Editor: Gabrielle Iglesias Layout and graphic artist: Lowil Fred EspadaADPC 2010*, available at: http://repo.floodalliance.net/jspui/bitstream/44111/1346/1/Planning%20for%20 Disaster%20Risk%20Reduction.pdf (accessed 15 July 2017).

Bacanovic, V. (2014), *Gender Analysis of the Impact of the 2014 Floods in Serbia*, OSCE Mission in Serbia, Belgrade.

Bijelić, M. and Lazarević, M. (2015), *Resilient Financing: The Economic Costs of Natural Disasters a Case Study of 2014 Serbia Floods*, UNDP, Belgrade.

Beck, U. (2001), *Rizično društvo – u Susret Novoj Moderni*, Filip Višnjić, Beograd.

Brown, O. (2008), *Migration and Climate Change*, International Organization for Migration (IOM), Geneva, available at: www.iom.cz/files/Migration_and_Climate_Change_-_IOM_Migration_ Research_Series_No_31.pdf (accessed 15 December 2016).

Crncevic, T. (2013), "Planiranje i zastita prirode, prirodnih vrednosti i predela u kontekstu klimatskih promena u Republici Srbiji", *Posebna Izdanja Br.72*, Institut za arhitekturu i urbanizam Srbije, JP "Sluzbeni glasnik", Beograd.

Crncevic, T., Maric, I. and Josimovic, B. (2011), "Strategic environmental assessment and climate change in the republic of Serbia – support to development and adjustment process", *Spatium 26*, pp. 14-19.

Crncevic, T., Dzelebdzic, O. and Milijic, S. (2016), "Planning and climate change: a case study on the spatial plan of the Danube corridor through Serbia", in Leal Filho, W., Adamson, K., Dunk, R.M., Azeiteiro, U.M., Illingworth, S. and Alves, F. (Eds), *Implementing Climate Change Adaptation in Cities and Communities*, Springer, pp. 161-177.

Dragicevic, S., Filipovic, D., Kostadinov, S., Ristic, R., Novkovic, I., Zivkovic, N., Andjelkovic, G., Abolmasov, B., Secerov, V. and Djurdjic, S. (2011), "Natural hazard assessment for land-use planning in Serbia", *International Journal Environmental Research*, Vol. 5 No. 2, pp. 371-380.

International Federation of Red Cross and Red Crescent Societies (IFRC) (2015), *Annual Report 2015*, available at: http://media.ifrc.org/ifrc/wp-content/uploads/sites/5/2016/11/IFRC-Annual-Report-2015-EN_LR.pdf (accessed 10 January 2017).

International Federation of Red Cross and Red Crescent Societies (IFRC) (2016), *World Disasters Report Resilience: Saving Lives Today, Investing for Tomorrow*, available at: www.ifrc.org/Global/ Documents/Secretariat/201610/WDR%202016-FINAL_web.pdf (accessed 15 March 2017).

Internal Displacement Monitoring Centre (IDMC), Norwegian Refugee Council (NRC) (2016), *GRID 2016: Global Report on INTERNAL DISPLACEMENT*, Internal Displacement Monitoring Centre, available at: www.internal-displacement.org/assets/publications/2016/2016-global-report-internal-displacement-IDMC.pdf (accessed 10 January 2017).

Intergovernmental Panel on Climate Change (IPCC) (2014a), "Climate change 2014: synthesis report", in Core Writing Team, Pachauri, R.K. and Meyer, LA. (Eds), Contribution of working groups I, II and III to the fifth assessment report of the Intergovernmental Panel on Climate Change, IPCC, Geneva.

Intergovernmental Panel on Climate Change (IPCC) (2014b), "Climate change 2014: synthesis report", in *Climate Change 2014 Synthesis Report Summary for Policymakers*, available at: www.ipcc.ch/pdf/assessment-report/ar5/syr/AR5_SYR_FINAL_SPM.pdf (accessed 10 January 2016).

Krstic, G., Arandarenko, M., Nojkovic, A., Vladisavljević, M. and Petrović, M. (2010), "Polozaj ranjivih grupa na trzistu rada, Beograd: Fondacija za unapredjenje ekonomije", available at: www.inkluzija.gov.rs/wp-content/uploads/2010/04/Polozaj-ranjivih-grupa-na-trzistu-rada-FREN.pdf (accessed 10 November 2016).

Miladinovic, S. (2016), "Forced migrations and their social implications", *Book of Abstracts from the International Conference on Modern Migrations and Social Development: Interdisciplinary Perspective*, Faculty of Philosophy, Belgrade, October 21-22, 2016.

Nicholls, R.J. and Lowe, J. (2004), "Benefits of migration of climate change for coastal areas", *Global Environmental Change*, Vol. 14 No. 3, pp. 229-244.

Orlovic-Lovren, V. (2014), *Report on Needs Assessment in Public Education in DRR at the Local Level in Serbia*, UNDP Serbia, Belgrade.

Orlovic-Lovren, V. and Pejatovic, A. (2015), "The role of adult educators in fostering learning for community resilience", in Krasovec, S.J. and Stefanc, D. (Eds), *Conference Proceedings on Perspectives on Community Practices: Living and Learning in Community*, ESREA, Ljubljana.

Official Gazette of the Republic of Serbia (2010), *Zakon o Prostornom Planu Republike Srbije Br. 88/2010*, Official Gazette of the Republic of Serbia.

Official Gazette of the Republic of Serbia (2011), *Nacionalna Strategija Zaštite i Spasavanja u Vanrednim Situacijama, Br. 86/11*, Official Gazette of the Republic of Serbia.

Official Gazette of the Republic of Serbia (2012), *National Strategy for Sustainable Use of Natural Resources and Goods*, Official Gazette RS 33/12, Official Gazette of the Republic of Serbia.

Petovar, K. and Jokic, V. (2011), "The right of servitude between public interest and undisturbed use of private property", *SPATIUM International Review*, Vol. 26, pp. 7-13.

Raleigh, C. and Jordan, L. (2010), "Climate change and migration: emerging patterns in the developing world", Mearns, R. and Norton, A. (Eds), *Social Dimensions of Climate Change: Equity and Vulnerability in a Warming World*, The World Bank, Washington, DC, pp. 103-132.

Raleigh, C., Jordan, L. and Salehyan, I. (2008), "Assessing the impact of climate change on migration and conflict", *Social Development*, The World Bank, Washington, DC, available at: http://siteresources.worldbank.org/EXTSOCIALDEVELOPMENT/Resources/SDCCWorkingPaper_MigrationandConflict.pdf (accessed 10 December 2016).

Republika Srbija (RS) (2017), "Drugi izvestaj Republike Srbije prema Okvirnoj konvenciji Ujedinjenih nacija o promeni klime", available at: www.klimatskepromene.rs/uploads/useruploads/Documents/SNC_na-misljenje.pdf (accessed 10 February 2017).

Stefanovic, M. and Gavrilovic, Z. (2014), "Lokalna zajednica i problematika bujicnih poplava, Prirucnik za lokalnu zajednicu i organiyacije civilnog drustva", *Organizacija za Evropsku Bezbednost i Saradnju*, Misija u Srbiji, Beograd, Fiducia 011:print.

Steffen, W., Grinewald, J., Crutzen, P. and Mc Neill, J. (2011), "The Anthropocene: conceptual and historical perspectives", *Philosophical Transactions of the Royal Society A*, Vol. 369 No. 1938, pp. 842-867, doi: 10.1098/rsta.2010.0327, available at: http://rsta.royalsocietypublishing.org (accessed 23 November 2016).

United Nations (2015), *Resolution Adopted by the General Assembly on 25 September 2015*, available at: www.un.org/ga/search/view_doc.asp?symbol=A/RES/70/1&Lang=E (accessed 10 January 2017).

United Nations (UN) Serbia, European Commission (EC), World Bank Group (WBG) (2014), *Poplave u Srbiji 2014*, Beograd, available at: www.obnova.gov.rs/uploads/useruploads/Documents/ Izvestaj-o-proceni-potreba-za-oporavak-i-obnovu-posledica-poplava.pdf (accessed 10 November 2016).

United Nations Development Programme (UNDP) (2016), *Human Development Report 2016: Risk Proofing the Western Balkans: Empowering People to Prevent Disasters*, UNDP, Sarajevo, available at: http://hdr.undp.org/sites/default/files/risk_proofing_the_western_balkans.pdf

UNHCR Serbia update (2017), available at: http://reliefweb.int/report/serbia/unhcr-serbia (accessed 1 March 2017).

UNISDR (2015), *Sendai Framework for Disaster Risk Reduction 2015-2030*, available at: www.unisdr. org/files/43291_sendaiframeworkfordrren.pdf (accessed 10 November 2016).

Upadhyay, H., Kelman, I., Lingaraj, G.J., Mishra, A., Shreve, C. and Stojanov, R. (2015), "Conceptualizing and contextualizing research and policy for links between climate change and migration", *International Journal of Climate Change Strategies and Management*, Vol. 7 No. 3, pp. 394-417.

World Meteorological Organization (WMO), United Nations Environment Programme (UNEP), Intergovernmental Panel on Climate Change (IPCC) (1990), *Climate Change, The IPPC Scientific Assessment*, available at: www.ipcc.ch/publications_and_data/publications_ipcc_first_ assessment_1990_wg1.shtml (accessed 10 January 2016).

Yonetani, M. (2017), *Briefing Paper Positioned for Action, Displacement in the Sendai Framework for Disaster Risk Reduction*, International displacement monitoring center (IDMC), Norwegian Refugee Council (NRC), available at: www.internal-isplacement.org/assets/publications/2017/ 20170216-idmc-briefing-paper-drr.pdf (accessed 1 July 2017).

Further reading

Republika Srbija, Ministarstvo zivotne sredine i prostornog planiranja (2011), *Strategija Bioloske Raznovrsnosti Republike Srbije za period 2011-2018*, available at: www.zzps.rs/novo/kontent/ stranicy/propisi_strategije/strategija_bioloske_raznovrsnosti.pdf (accessed 10 November 2016).

Corresponding author
Tijana Crnčević can be contacted at: tijana@iaus.ac.rs

9

Meteorological drought assessment in north east highlands of Ethiopia

Yimer Mohammed
Hawassa University, Hawassa, Ethiopia and Dilla University, Dilla, Ethiopia

Fantaw Yimer and Menfese Tadesse
Hawassa University, Hawassa, Ethiopia, and

Kindie Tesfaye
*International Maize and Wheat Improvement Center (CIMMYT),
Addis Ababa, Ethiopia*

Abstract

Purpose – The purpose of this paper is to investigate the patterns and trends of drought incidence in north east highlands of Ethiopia using monthly rainfall record for the period 1984-2014.

Design/methodology/approach – Standard precipitation index and Mann – Kendal test were used to analyze drought incident and trends of drought occurrences, respectively. The spatial extent of droughts in the study area has been interpolated by inverse distance weighted method using the spatial analyst tool of ArcGIS.

Findings – Most of the studied stations experienced drought episodes in 1984, 1987/1988, 1992/1993, 1999, 2003/2004 and 2007/2008 which were among the worst drought years in the history of Ethiopia. The year 1984 was the most drastic and distinct-wide extreme drought episode in all studied stations. The Mann–Kendal test shows an increasing tendencies of drought at three-month (spring) timescale at all stations though significant ($p < 0.05$) only at Mekaneselam and decreasing tendencies at three-month (summer) and 12-month timescales at all stations. The frequency of total drought was the highest in central and north parts of the region in all study seasons.

Originality/value – This detail drought characterization can be used as bench mark to take comprehensive drought management measures such as early warning system, preparation and contingency planning, climate change adaptation programs.

Keywords Drought, Trends, SPI, Extremes, Mann–Kendall

1. Introduction

Drought is a recurrent climate phenomenon which occurs in most parts of the world, with varying frequency, severity and duration (Wilhite, 1993; Shatanawi *et al.*, 2013). It is difficult to determine the onset and ending time of a drought. It develops slowly, and its impact may remain for years after termination of the event (Morid *et al.*, 2006). There is no universally accepted definition of drought that applies to all circumstances. Perhaps most definitions are based on the deficiency of rainfall resulting in water shortage for some activities related to use of water (Wilhite, 1993; Wilhite and Glantz, 1985). Morid *et al.* (2006), Karavitis (1999) and Szinell *et al.* (1998) defined drought as "the state of adverse and wide spread hydrological, environmental, social and economic impacts due to less than generally anticipated water quantities".

Droughts are one of the highest natural disasters globally having major impacts on environmental, economic and social conditions (Paulo *et al.*, 2012; Morid *et al.*, 2006). The impact of drought is governed by the magnitude, duration, frequency and spatial extent of the rainfall deficit (Degefu and Bewket, 2013; Zargar *et al.*, 2011). Magnitude refers to the amount of rainfall or water storage deficit at a particular place and specific time. Drought magnitude is categorized, in most commonly used indices, into mild, moderate, severe and extreme. Frequency/return period/is the average time between drought events and duration refers to the length of time that a given drought event stays. The spatial coverage refers to the areal extent of a specific area affected by a given drought incidence. Usually severe and extreme drought episodes cover wider areas, while mild and moderate drought episodes tend to affect localized areas (Degefu and Bewket, 2013).

Scientific literature has commonly classified droughts into four main categories as meteorological, agricultural, hydrological and socio-economic (Wilhite and Glantz, 1985; Tallaksen and Van Lanen, 2004; Mishra and Singh, 2010). For the purpose of this study, meteorological drought was adopted where rainfall is commonly used for drought analysis. Droughts over a region can be characterized using different indices, which all use rainfall either alone or in combination with other meteorological elements (Zargar *et al.*, 2011; Edossa *et al.*, 2010).

Although drought is a natural occurring recurrent extreme event (Wilhite, 1993; Shatanawi *et al.*, 2013), various empirical and modeling studies proved that climate change is very likely to increase the magnitude, frequency and duration of droughts over some parts of the world in the coming decades (IPCC II, 2014, Degefu and Bewket, 2013).

There were severe droughts which have had a substantial impact on the socio-economic and environmental condition of Ethiopia at different times and scales (Table I). For instance, although the recorded history of drought in Ethiopia dates back to 250 BC, its frequency has increased over the past few decades and still have been the hallmarks of the country (Meze-Hausken, 2000; Deressa *et al.*, 2010). From 1950s to1980s, droughts occurred on average once per decade, and recently, it occurred once every three years in Ethiopia (Block, 2008). On the other hand, World Bank (2006) reported that 16 drought events were experienced during 1980-2004 which makes Ethiopia the most drought-affected country. The year leading to June 2011 has been claimed to be the driest in 60 years in some regions of Somalia, Northern Kenya and Southern Ethiopia (USAID/FEWS 2011). Ethiopia faced one of the worst droughts the country has seen in decades with over 10.2 million people in need of food aid (UN OCHA, 2015). Triggered by El Niño, the rainy seasons in 2015 failed so that the drought brought significant impact by limiting agricultural production, straining livelihoods and exacerbating food insecurity among poor and vulnerable households.

From 1986 until 2013, a total of nine El Niño events had occurred. The magnitude of seven of the El Niño years (1986/1987, 1991/1992, 1994/1995, 2002/2003, 2004/2005, 2006/2007, 2009/2010) were either moderate or weak, while the magnitude of two of them (1987/1988 and 1997/1998) were strong (FAO, 2014). There is a remarkable correspondence between annual rainfall in Ethiopia and ENSO events (Haile, 1988).

Table I. Major drought years and their effects in different regions of ethiopia for the last 50 years

Year	Region	Impacts
1964-1966	Tigray and Wollo	About 1.5 million people were affected and 300,000 livestock died
1972-1973	Tigray and Wollo	Death of about 200,000 people and 30% of livestock population in the area
1978-1979	Southern Ethiopia	1.4 million people were affected
1983-1984	All regions	8 million people affected, 1 million people died
1982	Northern Ethiopia	2 million people were affected
1987-1988	All regions	7 million people were affected
1991-1992	North, east & south Ethiopia	4 million people were affected
1993-1994	Tigray and Wollo	7.6 million people were affected
2000	All regions	About 10.5 million people were affected
2002-2003	All regions	About 13 million people were affected; 1.4 million livestock died
2006	Southern Ethiopia (Borena)	About 7.4 million people affected; 247,000 livestock died
2008	Southern Ethiopia (Borena)	About 26,000 livestock died
2008-2009	All regions	About 5 million people were affected
2011	South-central, southeastern, and eastern parts of Ethiopia	About 4.5 million were affected
2015-2016	Northern, Southern and Eastern Ethiopia	About 10.2 million people were affected

Sources: Compiled from Degefu (1987), Meze-Hausken (2000), FAO (2003), Segele and Lamb (2005), Amsalu and Adem (2009), Deressa *et al.* (2010), Famine Early Warning Systems Network (FEWS NET) (2011), Viste *et al.* (2012) and FDRE (2016)

The analysis and forecasting of extreme climatic events has become increasingly relevant to make planning effective. Although drought is becoming the most common and damaging natural hazard in Ethiopia, there were no as such detail studies conducted about its magnitude, frequency and spatial extent in different regions of Ethiopia (Degefu and Bewket, 2013). The same is true in this study area. Hence, to reduce the damages from drought, it is crucial to characterize drought. Drought characterization at regional and local scales has significant implications for drought management such as early warning system (farmers could be warned before the advent of drought as to when, what and where to cultivate and when to sell their animals as well as how to conserve water resources and food), preparation and contingency planning (the government is working toward reorganizing its resources *before* the impact of drought is felt, climate change adaptation programs (introduction of off-farm activities and drought resistant crop and animal varieties).

Different indices have been developed through time to quantify the magnitude of meteorological drought. Most of these indices are based on direct observed measurement of climatic variables such as rainfall, evapotranspiration and temperature (Steinemann *et al.*, 2005). Some of the widely used indices include the Palmer Drought Severity Index (PDSI) (Palmer, 1965), the Percent of Normal, and the Deciles approach (Gibbs and Maher, 1967), the Standard Precipitation Index (SPI) (McKee *et al.*, 1993) and Standard Precipitation Evapotranspiration Index (SPEI) (Vicente-Serrano *et al.*, 2010).

Each index has its own strength and weakness. For example, PDSI is considered as useful in monitoring drought at a regional scale and allows comparisons over relatively large zones (Steinemann *et al.*, 2005). It also uses water balance models to consider the effects of temperature and rainfall. However, it does not work well in mountainous areas where there are extremes in rainfall or runoff variables (Mpelasoka *et al.*, 2008; Ellis *et al.*, 2010).

Moreover PDSI requires large amount of climate and soil moisture data, and they are not simple to calculate (Degefu and Bewket, 2013; Morid *et al.*, 2006). A brief description of these and other indices was given in Morid *et al.* (2006).

The choice of indices for drought monitoring should be based on the quantity and quality of available climate data, purpose of the study, computational simplicity and its ability to consistently detect spatial and temporal variations of a drought event (Morid *et al.*, 2006). Keyantash and Dracup (2002) also indicated that drought indices must be statistically robust and easily calculated, and have a clear and comprehensible calculation procedure.

In this study, SPI, which is currently used widely for its multiple advantages, has been used. The SPI was proposed by McKee *et al.* (1993) and has been used frequently during the past two decades (Hirschi *et al.*, 2011; Vicente-Serrano *et al.*, 2015). The robustness of SPI over the other drought indices has been reported in many studies (Vicente-Serrano *et al.*, 2015; Viste *et al.*, 2012; WMO, 2012; Hirschi *et al.*, 2011; Hayes *et al.*, 1999; Guttman, 1998; Mpelasoka *et al.*, 2008). In addition to its simplicity and ease of calculation, in regions like Ethiopia, where the access to data is limited, there are good reasons for choosing a rainfall-based drought measure (SPI) (El Kenawy *et al.*, 2016; Degefu and Bewket, 2013; Viste *et al.*, 2012). Details of SPI computation can be accessed in the user guides of SPI (WMO 2012).

The objective of this paper is therefore to monitor the magnitude, duration, frequency, spatial variability and trends of drought incidence in South Wollo, north east highlands of Ethiopia.

2. Materials and methods

2.1 Description of the study area

South Wollo zone is located in the north east part of Ethiopia lying between 10°12′N and 11° 40′N and 38°30′E and 40°05′E. Its zonal capital, Dessie, is found 400 kms North of Addis Ababa. It is one of the drought-prone and aid-dependent areas in Amhara Regional State (Bewket and Conway, 2007). The study area covers a total area of 17053.45 km^2 with 18 rural and 2 urban districts.

South Wollo is characterized by diverse topographic features in which high mountainous and deeply incised canyons and gorges, valleys and plateaus with steep slopes dominate its most parts (Coltorti *et al.*, 2007). The elevation ranged from the dry plains at 1,000 m altitude in the east to the high peaks above 3,500 m altitude in the west. However high-land areas ranging between 1,500 and 3,500 m altitude are the dominating feature of South Wollo (Rosell and Holmer, 2007). Because of the diverse topography, the study area experiences different climatic conditions that range from hot and arid lowlands in the eastern part to cold and humid highlands in the western part.

The mean annual temperature varies from less than 5°C in the western highlands to 22°C in the eastern lowlands. The annual rainfall varies between less than 1,000 mm in the western part to more than 1,200 mm in the eastern part. Bimodal rainfall pattern with shorter spring (March-May) and longer summer (June-August) characterizes the study area which leads to two harvest periods. The spring season is very influential for mid to high latitude areas, whereas the mid lands to lowland areas mainly depend on the summer rain. Except a few pocket areas where small-scale irrigation is practiced, crop production is rainfed.

As the topography is very rugged, soil erosion was a critical problem in the study area. It is highly degraded and deforested in terms of indigenous trees but does have considerable eucalyptus plantations. Alpine species unique to extreme highland areas are found in western highland parts of the study area. Local people in the area are engaged in subsistence agriculture for their livelihood. Unlike other areas of East Africa where highlands generally are the most food secure parts of a country, the opposite often is the case in Ethiopia. In fact, the crowded, steep-sloped highlands above 2,000 metres including large parts of South Wollo, are among the country's most famine-prone areas (Little *et al.*, 2006) (Figure 1).

2.2 Data source
Monthly rainfall data from six stations for the period (1984-2014) were obtained from Ethiopian National Meteorological Agency (NMA) and used as an input variable to calculate meteorological drought. Initially, we collected rainfall data from 10 stations found in the study area. However, due to their missing data and shorter length of records, we only selected the six stations.

2.3 Standard Precipitation Index computation
SPI was designed to quantify the rainfall deficit for multiple timescales in the studied stations. The SPI is a z-score and represents the drought event departure from the mean, expressed in standard deviation units. SPI is a normalized index in time and space. This feature allows comparisons of SPI values among different locations.

Although SPI can be calculated from 1 month up to 72 months, 1-24 months is the best practical range of application (Guttman, 1999; WMO, 2012). We, therefore, computed the SPI values at two time-scales, i.e. 3 months (SPI-3) and 12 months or annual (SPI-12). The SPI-3 was used to assess droughts during spring (belg) and summer (kiremt) seasons which represent the shorter and longer rain seasons, respectively, and SPI-12 was used to assess the annual drought. Positive SPI values indicated greater than mean rainfall and negative values indicated less than mean rainfall.

For each month of the calendar year, new data series were created, with the elements equal to corresponding rainfall moving sums (Degefu and Bewket, 2013). Then, the SPI value provides a comparison of the rainfall over a specific period with the rainfall totals

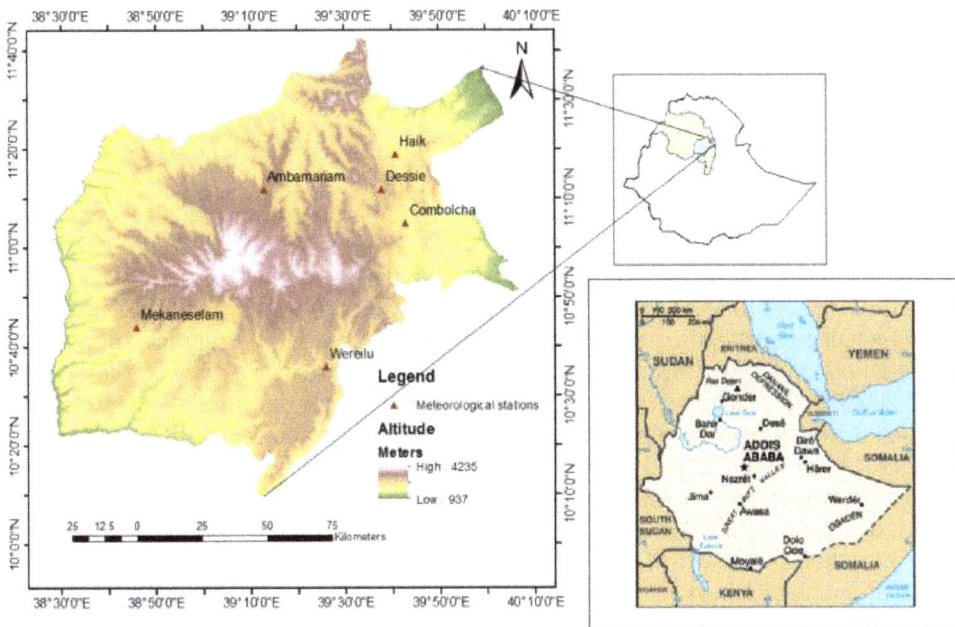

Figure 1. Location map of South Wollo and meteorological stations used in this study

from the same period for all the years included in the historical record (Shahid, 2008). It is essentially a seasonally normalized, backwards-looking moving average of precipitation. For example, the three-month SPI calculated for August 2005 used the total rainfall of June-July-August 2005 and compared with the June-August rainfall totals of all the years considered in the study. Similarly, the 12-month SPI for December 2005 used the rainfall total for January 2005 to December 2005.

Conceptually, SPI is equivalent to the Z-score used in statistics, and it is calculated as:

$$SPI_{ij} = \frac{X_{ij} - \mu_{ij}}{\alpha_{ij}}$$

where SPI_{ij} is the SPI of the ith month at the jth timescale, X_{ij} is rainfall total for the ith month at the jth time scale, μ_{ij} and α_{ij} are the long-term mean and standard deviation associated with the ith month at the jth time scale, respectively.

In this study, SPI values were produced for each month of the year using SPI_SL.6.exe program developed by Colorado Climate Center (available at http://ulysses.atmos.colostate.edu/SPI.html).

This program classified drought events in four magnitude classes, i.e. if SPI ranges between 0 and -0.99, it is mild drought; -1.00 and -1.49, it is moderate drought; -1.50 and -1.99, it is severe drought; and extreme drought if SPI is -2.00 or less (McKee et al., 1993; WMO, 2012). Each drought event has a duration defined by its beginning and end.

2.4 Methods for smoothing time series data and trend detection

The rank-based nonparametric Mann–Kendall (MK) test (Mann, 1945; Kendall, 1975) has been commonly used to assess the significance of monotonic trends in hydro-meteorological time series. Another rank-based nonparametric test, the Spearman's rho (SR) test (Lehmann, 1975; Sneyers, 1990), has sometimes been applied to detect trends in hydrological data. The study of Yue et al. (2002) and Viste et al. (2012) noted that these two tests have almost the same power to identify trends in time series. In comparison to the parametric t test, the common use of the nonparametric tests is mainly due to the consideration that they are more suitable for the situations of non-normal data, incomplete data, and missing data problems, which frequently occur in hydro-meteorological studies. MK test simply calculates whether the variable is increasing or decreasing with time (Collins, 2009).

MK test, used by many researchers for trend detection due to its robustness for non-normally distributed data, was applied in this study to assess trends in the time series data (Kendall, 1975; Mann, 1945). However, the MK test requires the time series data to be serially independent (Petrow and Merz, 2009). The standard p-values obtained from it are based on an assumption of independence between observations. The presence of a significant positive serial correlation tends to overestimate the significance of the trend (Petrow and Merz. 2009; Yue et al., 2002). Thus, the time series data should be tested for serial correlation and should be corrected if there is any significant autocorrelation (Petrow and Merz, 2009; Burn et al., 2010) before subjected the data to the MK trend test. To improve the performance of the test (MK), the most widely used procedure of the trend-free pre-whitening (TFPW) method can be applied to remove serial correlation from the time series, and hence to eliminate the effect of serial correlation on the MK test (Yue et al., 2002, 2003; Petrow and Merz, 2009). In this study, we found serial correlation of six out of eighteen time scales. However, only Ambamariam and Wereilu at 12-month timescale showed significant autocorrelation ($p = 0.59$ and 0.55), respectively. TFPW method was applied to remove the autocorrelation of these two stations following the works of Petrow and Merz (2009), Yue *et al.* (2002). Finally, we run the MK test. For detailed MK test statistics, see Kendall (1975) and Mann (1945) and the MK test manual prepared by Finnish Meteorological Institute (2002).

2.5 Mapping spatial distribution of drought incidences

The output from the SPI program is used as an input to ArcGIS to generate drought severity maps for the study area at 3 and 12-month time scales. To assess the spatial extent of droughts in the study area, SPI time series values of each meteorological station have been interpolated by Inverse Distance Weighted (IDW) method using the Spatial Analyst tool of ArcGIS. The IDW method gives better representation for interpolation of rainfall distribution over heterogeneous topographic terrain (Tagel *et al.*, 2011).

3. Results

The results of the study are discussed under three sub-topics:

(1)   frequency and magnitude of drought events;

(2)   trends of occurrences of drought events;

(3)   spatial patterns of drought events.

3.1 Magnitude and frequency of drought events

Rainfall is bimodal in the study area (Figure 2). The contribution of summer (long rain period) rainfall to the average annual rainfall across stations was very high ranging from 65.7 per cent at Ambamariam to 53.56 per cent at Dessie. The number of drought months with different magnitude classes at short and long timescales (3 and 12-months) is shown on Figure 3. Drought months for 3 months (belg and kiremt rainfall) and 12 months (annual rainfall) were calculated using SPI. Drought frequency, in this study, was measured by the number of years which experienced negative SPI values in the total time series of 30 years.

3.1.1 Spring season. The total number of drought events with mild, moderate, severe and extreme intensities computed at three-month timescale (March-May) was accounted for 46 per cent in spring season in all stations except Combolcha and Ambamariam (not shown here). However, they had varied magnitude classes. Extreme magnitude droughts occurred for 2 months at Ambamariam, Combolcha and Dessie. As the analysis of three-month timescale (March-May) showed, the year 1988, 1999, 2004 and 2007-2009 were drought years across all studied stations. Except Haik and Wereilu which had severe drought magnitude,

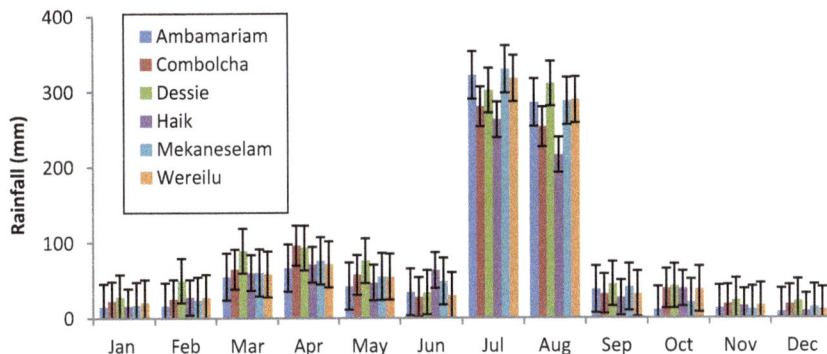

Figure 2. Bimodal rainfall distribution of the study area

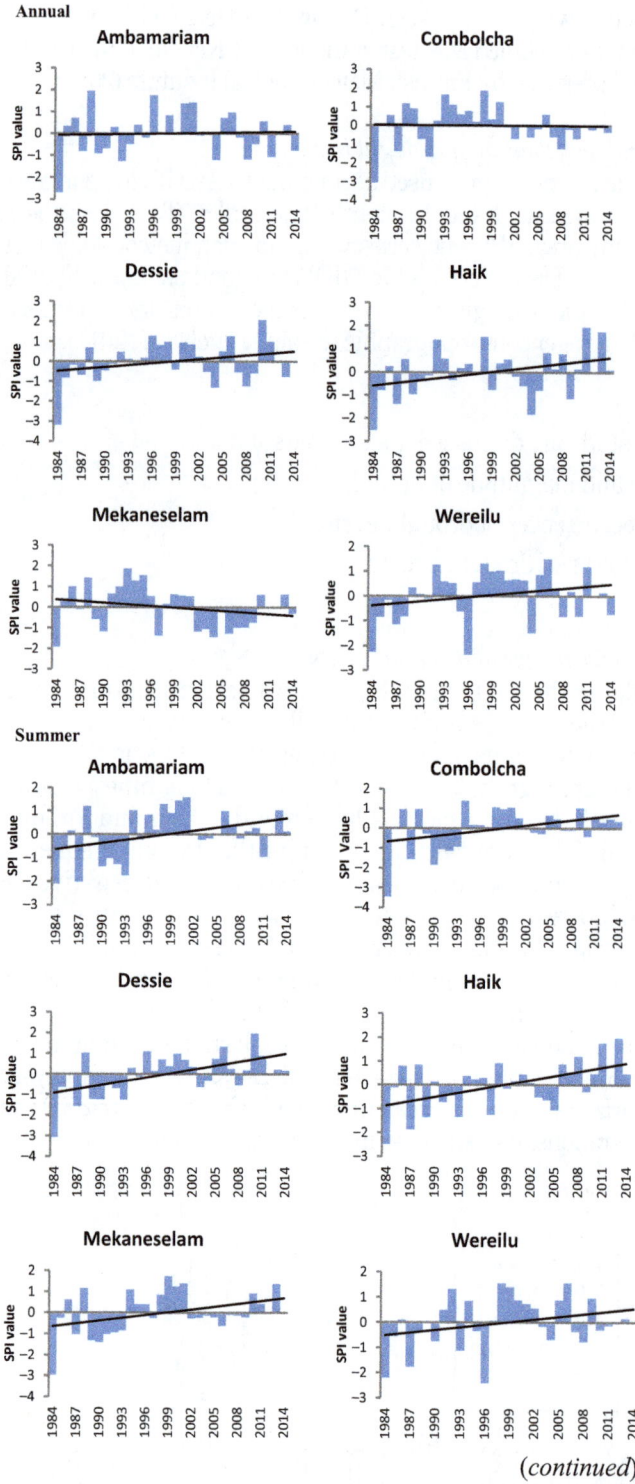

(*continued*)

Figure 3. Magnitude and frequency of SPI values at annual, summer and spring time scales

Spring

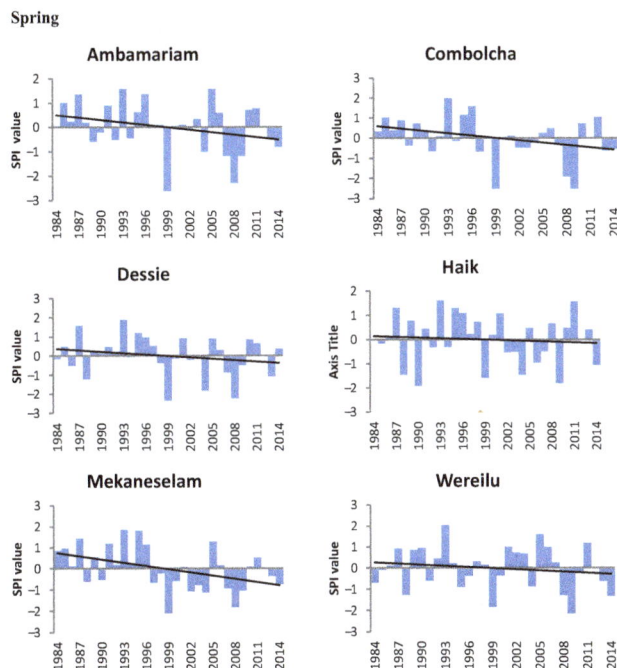

Figure 3.

extreme drought was recorded in 1999 in all stations ranging from −2.09 to −2.61. In the remaining drought years, the drought magnitude varied from mild to extreme in which Ambamariam and Dessie had extreme drought in 2008 and Combolcha and Wereilu experienced extreme drought in 2009.

3.1.2 Summer season. The total number of drought events at the three-month timescale (June-August) in the entire period of analysis was found between 12 months at Dessie and 18 months at Mekaneselam, respectively. In addition to 1984 drought year, Ambamariam had extreme (SPI value −2) drought at this timescale (June-August) in 1987 and Wereilu with SPI value of −2.39 in 1996. The worst drought was recorded in 1984 which had substantial impact on the region. Even though its magnitude varied, drought occurred across the study area at this timescale in 1984, 1987, 1992-1993, 2004 and 2008. The frequency of occurrences of drought at this season was the highest at Mekaneselam (60 per cent) and the lowest at Dessie (53 per cent) implying that Mekaneselam experienced one drought episode in almost every two to three years.

3.1.3 Annual drought. The total number of drought months at 12 month timescale (January-December) was found between 12 at Wereilu and 16 at Ambamariam which constitutes 40 and 53 per cent of the total number of drought incidences in the study period, respectively. At this time scale, the highest frequency of extreme drought was recorded at Wereilu in 1984 and 1996 with SPI value of −2.24 and −2.33, respectively. The year 1984 was the worst drought across the region with SPI value ranging from −1.9 to −3.17. Although there was varying degree of severity, drought was recorded across the study region in 1984, 1987, 2004 and 2008-2009. The total number of moderate droughts at 12 month timescale was the highest at Ambamariam and Mekaneselam with 4 and 6 months, respectively. This study confirms that the frequency of occurrences of droughts at summer was highest at Ambamariam.

There is a remarkable correspondence between rainfall in Ethiopia and ENSO events (Haile, 1988). When past ENSO events are compared with drought and famine periods in Ethiopia, they show a remarkable association. Recent documented droughts of Ethiopia in 1987-1988, 1991-1992, 1993-1994, 2002-2003, 2008-2009, 2011, 2015-2016 were all strong El Nino years (UN OCHA, 2015; FAO, 2014). This suggests that Ethiopian intense and extended drought periods either coincide or follow El Nino events. In recent years, the meteorological droughts are increasing especially in spring (belg) season (Figure 3) in the study region, and they have strong correlation with the El Nino events. Forecasting the occurrences of El Nino events will provide insights to estimate drought occurrences and plan adaptation strategies to minimize the associated risks.

3.2 Trends of drought occurrences
Trends of drought occurrences for 3 months (spring and summer) and 12 months (annual) timescales were shown in Figure 4. SPI values of May, August and December were considered to represent the drought conditions from March-May (spring season), June-August (summer season) and January-December (annual rainfall), respectively.

The computed SPI values for 3 months timescale during spring revealed that the occurrence of negative rainfall anomalies or frequent droughts were observed in the 1980s and 2000s and positive rainfall anomalies in the 1990s at most of the meteorological stations. The SPI value for 3 months timescale during summer showed increasing trends, which indicates declining of occurrences of droughts from time to time across the study stations. The SPI value for 12 months (annual) timescale showed negative rainfall anomalies at 1980s, the beginning of 1990s and after the mid of 2000s, while positive rainfall anomalies were observed at the end of 1990s and beginning of 2000s in the studied period.

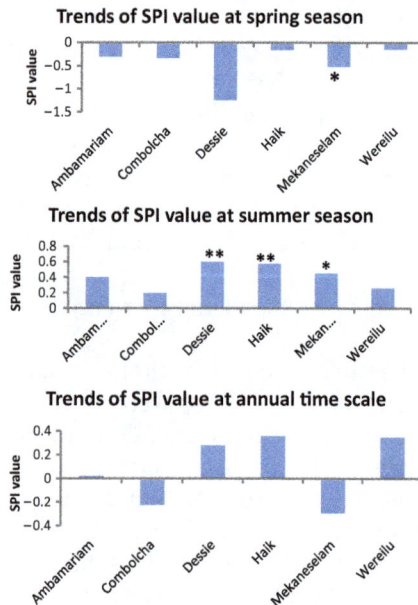

Notes: *Significant at $p < 0.05$ level;
**significant at $p < 0.01$ level

Figure 4. The Mann–Kendall's trend test for spring, summer and annual time scales

The MK trend test showed decreasing changes in SPI values in all stations suggesting increasing tendency of drought incidence at the three-month timescales during spring (Figure 4) However, significant trend ($p < 0.05$) was observed only at Mekaneselam station. On the other hand, the SPI values at the three-month timescale during summer showed increasing changes across stations, however, significant at Dessie, Haik and Mekaneselam ($p < 0.05$). Similar increasing trend of drought was also observed at 12-month timescale in all stations except Combolcha and Mekaneselam. However, the changes were not significant.

The trend analysis shows that there was no statistical evidence of any positive or negative trend of meteorological drought severity and frequency for the study area except Mekaneselam at the three-month timescale during spring and Dessie, Haik and Mekaneselam at the three-month timescale during summer. Although trends at all time-scales were not statistically significant, increasing tendencies of drought were observed during spring season and decreasing tendencies of drought during summer and annual scale in the study region.

3.3 Spatial patterns of drought occurrences
This study examined patterns of drought across the study region using time series (1984-2014) standardized precipitation index (SPI). The spatial patterns of drought at different time steps (three- and 12-months) were depicted in Figures 5(a)-(d) to 8(e1)-(e3). The

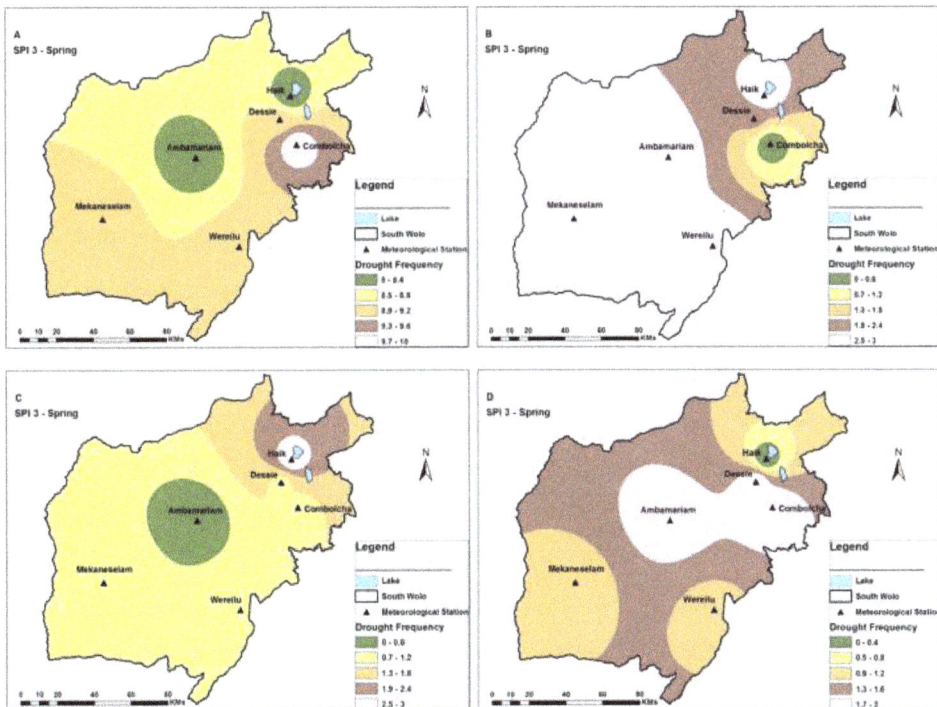

Figure 5. Spatial distribution of drought events at the three-month time scale in spring (March-May) season in South Wollo, north east highlands of Ethiopia: (A) mild drought, (B) moderate drought, (C) severe drought, (D) extreme drought

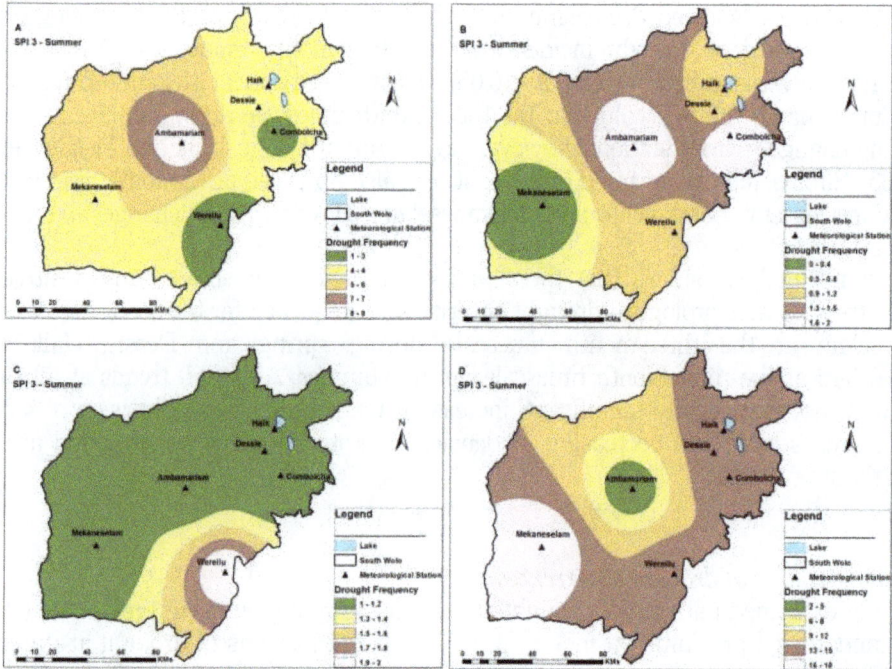

Figure 6. Spatial distribution of drought events at the three-month time scale in summer (June-August) season in South Wollo, north east highlands of Ethiopia: (A) mild drought, (B) moderate drought, (C) severe drought, (D) extreme drought

frequencies of drought for different magnitude classes and timescales show complex and local scale spatial patterns.

Except small pocket areas with severe droughts in eastern stations, the other stations were prone to moderate drought magnitude during spring. Extreme drought magnitude was also occurred at Ambamariam, Combolcha and Dessie in the same season. During summer, the western stations (Ambamariam, Mekaneselam and Wereilu) were highly exposed to high frequency of mild and moderate droughts, whereas eastern stations were less exposed to droughts. At annual timescale, extensive areas in the north east part of the region (Ambamariam, Combolcha and Dessie) also experienced high frequency of drought. Generally, the total frequency of drought was the highest at Mekaneselam at spring and summer and at Ambamariam at annual time scale.

4. Discussions

In this study, we presented a brief drought analysis at annual and seasonal time steps using SPI. We assessed the magnitude, frequency, trend, spatial pattern and probability of drought events over South Wollo from 1984 to 2014 at 3- and 12-month time scales. As the study area experiences bimodal rainfall (spring and summer) occurring between March to May and June to August respectively, we computed the SPI at three-month timescale for both seasons.

Results indicate that occurrences of drought in South Wollo showed temporal variation. The total number of drought events was higher in summer season than spring at Ambamariam, Combolcha and Mekaneselam. Summer is the main rainy season in the study

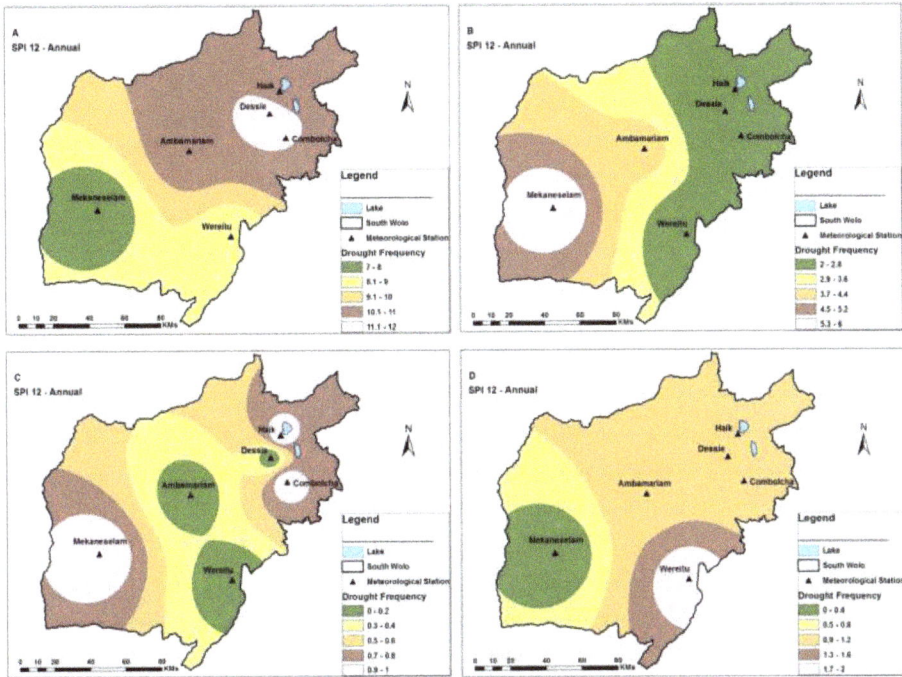

Figure 7. Spatial distribution of drought events at 12-month time scale in South Wollo, north east highlands of Ethiopia: (A) mild drought, (B) moderate drought, (C) severe drought, (D) extreme drought

area, contributing to more than 70 per cent of the annual rainfall (Viste *et al.*, 2012). Failing to rain at this season means failure of 85-95 per cent of the Ethiopian food crop which is produced in this season (Degefu, 1987).

Most of the studied stations experienced extreme magnitude droughts. The year 1984 was the most drastic and distinct-wide drought episode. All stations experienced extreme magnitude drought ranging from −2.18 to −3.44 at 3-month (summer) and from −1.9 to −3.03 at 12-month (annual) timescales. Except Wereilu and Haik (which had severe magnitude), extreme magnitude droughts were also observed at all stations in 1999 during spring season. This finding agrees with the findings of Viste *et al.* (2012) who found driest conditions all over Ethiopia during spring season in 1999. Additional extreme magnitude drought also occurred during spring at Ambamariam and Dessie in 2008 with SPI values of −2.28 and −2.22, respectively. There were also extreme magnitude droughts at Combolcha and Wereilu during spring season in 2009 with SPI value of −2.5 and −2.14, respectively. At the three-month timescale during summer, Ambamariam and Wereilu experienced extreme magnitude droughts in 1987 and 1996 with SPI values of −2. and −2.39, respectively.

The frequency of occurrences of drought was the highest (60 and 53 per cent) at Mekaneselam and Ambamariam at 3 month during summer and 12-month annual timescale, respectively. This shows that Mekaneselam and Ambamariam were frequently stricken by drought compared to other studied stations in South Wollo.

The occurrences of extreme magnitude droughts have important implications on rainfed agriculture (Ellis *et al.*, 2010; Morid *et al.*, 2006). The impacts of 1984 drought greatly expanded food insecurity, malnutrition, devastated livelihoods and caused for the loss of

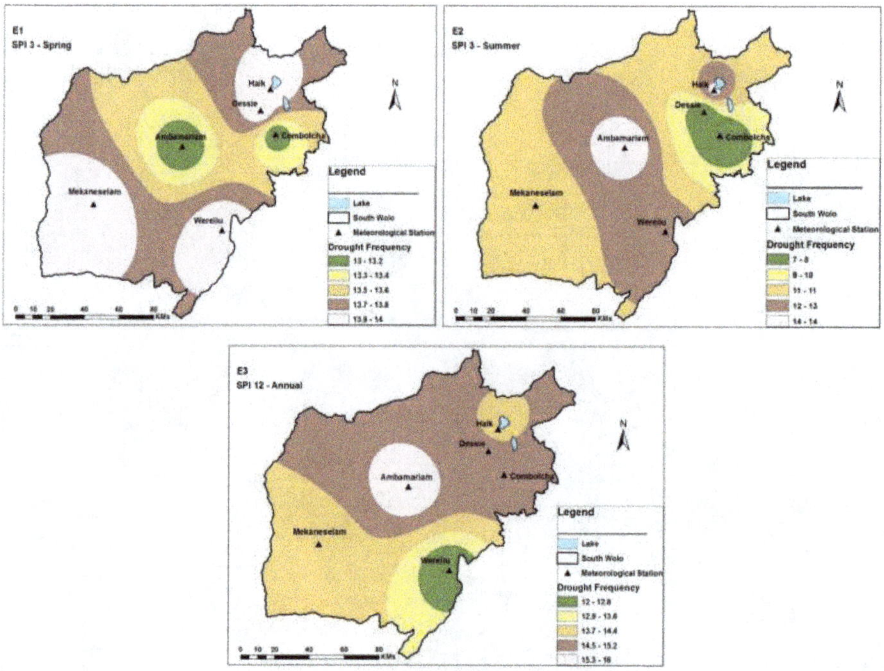

Figure 8. Spatial distribution of total drought events in, seasons and at annual timescales in South Wollo, north east highlands of Ethiopia: (E1) spring (March-May); (E2) summer (June-August); (E3) annual drought

millions of life in the region. This was also true all over Ethiopia. This finding agrees with previous research findings of (El Kenawy *et al.*, 2016; Mosaad, 2015; Bayissa *et al.*, 2015; Degefu and Bewket, 2013; Viste *et al.*, 2012; Tagel *et al.*, 2011; Edossa *et al.*, 2010; Bewket and Conway, 2007) who reported extreme drought in different parts of the country in the same period. Segele and Lamb (2005) also extensively demonstrated the severity of the 1984 drought over Ethiopia, particularly during the summer season.

The analysis revealed that in 1996 the SPI values at 3-month (June-August) and 12-month timescales at Wereilu and in 1987 at 3-month (June-August) at Ambamariam were unique to other stations which showed extreme droughts with SPI value of −2.39 and −2.33 at Wereilu, respectively, and −2.0 at Ambamariam. The analysis implies that drought magnitude can vary largely over space and this exceptional extreme drought shows a local character, where small area was affected, while others had less severe drought magnitude or even no drought. This finding is supported by the study of El Kenawy *et al.* (2016) who found similar drought occurrences in studying the changes in the frequency and severity of hydrological droughts over Ethiopia from 1960 to 2013.

The most striking characteristic of a drought is the change in its frequency as the time scale changes (Umran, 1999). On shorter time scales, droughts became more frequent but their duration was short. The opposite is true for longer time scales. Their frequency becomes less but remains for longer time. At the three-month scale, drought frequency increased but its duration decreased. This means that drought became more frequent at shorter time scales but stayed for shorter periods (Degefu and Bewket, 2013; Edossa *et al.*, 2010; Umran, 1999).

Generally, the temporal SPI analysis shows that most of the stations measured extreme, severe, moderate and mild drought episodes in the years 1984, 1987/1988, 1992/1993, 1999, 2003/2004 and 2007/08. Indeed, these years were among the worst drought years in the history of Ethiopia which agrees with previous findings of (Viste *et al.*, 2012, Tagel *et al.*, 2011 and Edossa *et al.*, 2010). The drought years identified by this SPI analysis for South Wollo are known for their substantial damage in terms of life and economic losses like other parts of Ethiopia which agrees with the findings of (Bayissa *et al.*, 2015; Tagel *et al.*, 2011). It is very important to remind that SPI used the monthly rainfall to analyze the drought episode in this study. However, the monthly rainfall data did not reflect the daily rainfall characteristics, such as the beginning and end time of dry spells which have a great implication on the effect of droughts especially for agricultural activities.

The MK test results indicate that SPI values were ranging from -0.14 to -0.51 Z-unit per decade at the three-month timescales during spring season suggesting increasing drought episodes. However, it was significant only at Mekaneselam ($p < 0.05$). On the other hand, the SPI values were increasing at the three-month (June-August) at all stations albeit the trends were statistically significant at Dessie, Haik and Mekaneselam. Generally, significant increasing tendencies of drought were observed during spring season and decreasing tendencies of drought during summer and annual scale in the study region.

5. Conclusions

In this study, a brief drought analysis was presented using SPI. It is a very important tool for quantifying drought and comparing its characteristics over time and space. We used SPI, in this study, to examine the magnitude and frequency, trend, patterns and probability of drought occurrences. Here, droughts occurrences were analyzed both at 3- and 12-month time scales, and the number of droughts at six stations are presented in Figure 3. Though almost all stations in the study region suffer from drought, it is important to consider that all of the stations did not experience well-defined drought episode during the same periods. In other words, temporal distribution and frequency of droughts varied markedly among each station.

Extreme droughts were more pronounced in stations where their altitude is above 3,000 masl during spring and in stations having altitude of less than 3,000 masl during summer. Similarly, stations found in western part of the study region were exposed to high frequency of severe and extreme droughts at annual timescale. The drought years in the study region identified by this SPI analysis were among the worst drought years in the history of Ethiopia. The year 1984, for example, was the most drastic and distinct-wide drought episode. Almost all stations experienced extreme magnitude drought at 3-month (summer) and 12-month timescales in the specified year.

Generally, the entire study area can be considered as drought prone area. Increasing tendencies of drought were observed during spring and decreasing tendencies at summer and annual timescales. The patterns of drought events in the study area are highly localized. Special attention (local-scale planning) should be given, while decision makers plan to effectively manage drought. The findings of this study have implications for drought management, early warning system, preparedness and contingency planning and climate change adaptation. In real sense, drought is a climatic event that cannot be prevented very easily, but interventions and preparedness to drought can help to cope with drought by developing more resilient ecosystems, improving resilience to recover from drought and taking various adaptation strategies like water harvesting, making irrigation system more efficient and a geographical shift of agricultural system.

References

Amsalu, A. and Adem, A. (2009), *Assessment of Climate Change-induced Hazards, Impacts and Responses in the Southern Lowlands of Ethiopia*, Forum for Social Studies, Addis Ababa.

Bayissa, A., Moges, A., Xuan, Y., Van Andel, J., Maskey, S., Solomatine, D., Griensven, V. and Tadesse, T. (2015), "Spatio-temporal assessment of meteorological drought under the influence of varying record length: the case of upper Blue Nile basin, Ethiopia", *Hydrological Sciences Journal*, Vol. 60 No. 11, pp. 1927-1942, doi: 10.1080/02626667.2015.

Bewket, W. and Conway, D. (2007), "A note on the temporal and spatial variability of rainfall in the drought-prone Amhara region of Ethiopia", *The International Journal of Climatology*, Vol. 27, pp. 1467-1477, doi: 10.1002/joc.1481.

Block, P.J. (2008), "Mitigating the effects of hydrologic variability in Ethiopia: an assessment of investments in agricultural and transportation infrastructure, energy and hydroclimatic forecasting", CPWF Working Paper 01: The CGIAR Challenge Program on Water and Food, Colombo, p. 53.

Burn, D.H., Sharif, M. and Zhang, K. (2010), "Detection of trends in hydrological extremes for Canadian watersheds", *Hydrological Processes*, Vol. 24 No. 13, pp. 1781-1790, doi: 10.1002/hyp.7625.

Collins, M.J. (2009), "Evidence for changing flood risk in new England since the late 20th century", *Jawra Journal of the American Water Resources Association)*, Vol. 45 No. 2, pp. 279-290, doi: 10.1111/j.1752-1688.2008.

Coltorti, M., Dramis, F. and Ollier, C.D. (2007), "Plantation surfaces in northern Ethiopia", *Geomorphology*, Vol. 89 Nos 3/4, pp. 287-296.

Degefu, W. (1987), "Some aspects of meteorological drought in Ethiopia", in Glantz, M.H. (Ed.), *Drought and Hunger in Africa: Denying Famine a Future*, Press Syndicate of the University of Cambridge, Cambridge, MA, pp. 223-236.

Degefu, M.A. and Bewket, W. (2013), "Trends and spatial patterns of drought incidence in the omo-ghibe river basin, Ethiopia", *Geografiska Annaler: Series A, Physical Geography*, doi: 10.1111/geoa.12080.

Deressa, T.T., Hassan, R.M. and Ringler, C. (2010), "Perception of and adaptation to climate change by farmers in the Nile Basin of Ethiopia", *Journal of Agricultural Science*, (Climate Change and Agriculture Paper), pp. 1-9.

El Kenawy, A.M., McCabe, M.F., Vicente-Serrano, S.M., López-Moreno, J.I. and Robaa, S.M. (2016), "Changes in the frequency and severity of hydrological droughts over Ethiopia from 1960 to 2013", *Cuadernos De Investigación Geográfica*, Vol. 42 No. 2, doi: 10.18172/cig.2925.

Edossa, D.C., Babel, M.S. and Gupta, A.D. (2010), "Drought analysis in the Awash river basin, Ethiopia", *Water Resources Management*, Vol. 24 No. 7, pp. 1441-1460, doi: 10.1007/s11269-009-9508-0.

Ellis, A.W., Goodrich, G.B. and Garfin, G.M. (2010), "A hydroclimatic index for examining patterns of drought in the Colorado river basin", *International Journal of Climatology*, Vol. 30, pp. 236-255, doi: 10.1002/joc.1882.

Famine Early Warning Systems Network (FEWS NET) (2011), *A Climate Trend Analysis of Ethiopia*, United States Geological Survey.

FAO (2014), "Understanding the drought impact of El Niño on the global agricultural areas: an assessment using FAO's Agricultural Stress Index (ASI)", ISBN 978-92-5-108671-1.

Gibbs, W.J. and Maher, J.V. (1967), *Rainfall Deciles as Drought Indicators, Bureau of Meteorology*, Bulletin no. 48, Melbourne.

Guttman, N.B. (1998), "Comparing the palmer drought index and the standardized rainfall index", *Journal of the American Water Resources Association*, Vol. 34 No. 1, pp. 113-121.

Guttman, N.B. (1999), "Accepting the standardized precipitation index: a calculation algorithm", *Jawra Journal of the American Water Resources Association*, Vol. 35 No. 2, pp. 311-322.

Haile, T. (1988), "Causes and characters of drought in Ethiopia", *Ethiopian Journal of Agricultural Sciences*, Vol. 10, pp. 1-2, 8-597.

Hayes, M., Wilhite, D.A., Svoboda, M. and Vanyarkho, O. (1999), "Monitoring the 1996 drought using the standardized rainfall index", *Bulletin of the American Meteorological Society*, Vol. 80 No. 3, pp. 429-438.

Hirschi, M., Seneviratne, S.I., Alexandrov, V., Boberg, F., Boroneant, C., Christensen, O.B. and Stepanek, P. (2011), "Observational evidence for soil-moisture impact on hot extremes in south eastern Europe", *Nature Geoscience*, Vol. 4 No. 1, pp. 17-21.

IPCC II (2014), "Climate change 2014; Impacts, adaptations and vulnerability", *Working Group II Contribution to the Fifth Assessment Report of Intergovernmental Panel on Climate Change*, Cambridge University Press, Cambridge, MA.

Karavitis, C.A. (1999), "Decision support systems for drought management strategies in metropolitan Athens", *Water International*, Vol. 24 No. 1, pp. 10-21.

Kendall, M.G. (1975), *Rank Correlation Methods*, Griffin, London.

Keyantash, J. and Dracup, J. (2002), "The quantification of drought: an evaluation of drought indices", *Bulletin of the American Meteorological Society*, Vol. 83 No. 8, pp. 1167-1180.

Lehmann, E.L. (1975), *Nonparametrics, Statistical Methods Based on Ranks*, Holden-Day, San Francisco, California.

Little, P.D., Stone, M.P., Mogues, T., Castro, A.P. and Negatu, W. (2006), "'Moving in place': drought and poverty dynamics in South Wollo, Ethiopia", *Journal of Development Studies*, Vol. 42 No. 2, pp. 200-225.

Mann, H.B. (1945), "Non-parametric tests against trend", *Econometrica*, Vol. 13 No. 3, pp. 245-259.

McKee, T.B., Doesken, N.J. and Kleist, J. (1993), "The relationship of drought frequency and duration to time scales", Eighth Conference on Applied Climatology, *American Meteorological Society*, *Anaheim, CA*, 17-22 January, pp. 179-184.

Meze-Hausken, E. (2000), "Migration caused by climate change: how vulnerable are people in Dryland areas? A case-study in northern Ethiopia", *Mitigation and Adaptation Strategies for Global Change*, Vol. 5, pp. 379-406.

Mishra, A.K. and Singh, V.P. (2010), "A review of drought concepts", *Journal of Hydrology*, Vol. 391 No. 1, pp. 202-216.

Morid, S., Smakhtinb, V. and Moghaddasi, M. (2006), "Comparison of seven meteorological indices for drought monitoring in iran", *International Journal of Climatology*, Vol. 26 No. 7, pp. 971-985, doi: 10.1002/joc.1264.

Mosaad, K. (2015), "Forecasting of meteorological drought using hidden Markov model (case study: the upper blue Nile river basin, Ethiopia)", *Ain Shams Engineering Journal*, Vol. 7 No. 1, pp. 47-56.

Mpelasoka, F., Hennessy, K., Jones, R. and Bates, B. (2008), "Comparison of suitable drought indices for climate change impacts assessment over Australia towards resource management", *International Journal of Climatology*, Vol. 28 No. 10, pp. 1283-1292, doi: 10.1002/joc.1649.

Palmer, W.C. (1965), "Meteorological droughts", *U.S. Department of Commerce Weather Bureau Research Paper 45*, p. 58.

Paulo, A.A., Rosa, R.D. and Pereira, L.S. (2012), "Climate trends and behaviour of drought indices based on rainfall and evapotranspiration in Portugal", *Natural Hazards and Earth System Science)*, Vol. 12 No. 5, pp. 1481-1491, doi: 10.5194/nhess-12-1481-2012.

Petrow, T. and Merz, B. (2009), "Trends in flood magnitude, frequency and seasonality in Germany in the period 1951-2002", *Journal of Hydrology*, Vol. 371 Nos 1/4, pp. 129-141, doi: 10.1016/j.jhydrol.2009.03.024.

Rosell, S. and Holmer, B. (2007), "Rainfall change and its implications for Spring harvest in South Wollo, Ethiopia", *Geografiska Annaler*, Vol. 89 No. 4, pp. 287-299.

Segele, Z.T. and Lamb, P.J. (2005), "Characterization and variability of summer rainy season over Ethiopia", *Meteorology and Atmospheric Physics*, Vol. 89 No. 1, pp. 153-180.

Shahid, S. (2008), "Spatial and temporal characteristics of droughts in the western part of Bangladesh", *Hydrological Processes*, Vol. 22 No. 13, pp. 2235-2247, doi: 10.1002/hyp.6820.

Shatanawi, K., Rahbeh, M. and Shatanawi, M. (2013), "Characterizing, monitoring and forecasting of drought in Jordan river basin", *Journal of Water Resource and Protection*, Vol. 5 No. 12, pp. 1192-1202.

Sneyers, R. (1990), "On the statistical analysis of series of observations", *Technical Note no. 143*, WMO-no. 415, World Meteorological Organization, Geneva.

Steinemann, A.C., Hayes, M.J. and Cavalcanti, L.F.N. (2005), "Drought indicators and triggers", in Wilhite, D.A. (Ed.), *Drought and Water Crises: Science, Technology, and Management Issues*, CRC Press.

Szinell, C.S., Bussay, A. and Szentimrey, T. (1998), "Drought tendencies in Hungary", *International Journal of Climatology*, Vol. 24, pp. 1441-1460, doi: 10.1002/(SICI)1097- 0088(199804)18:5.

Tagel, G., Veen, A.V.D. and Maathuis, B. (2011), "Spatial and temporal assessment of drought in the Northern highlands of Ethiopia", *International Journal of Applied Earth Observation and Geoinformation*, Vol. 13, pp. 309-321.

Tallaksen, L.M. and Van Lanen, H.A.J. (2004), "Hydrological drought", *Processes and Estimation Methods for Streamflow and Groundwater, Developments in Water Science*, Vol. 48, p. 579.

Umran, K.A. (1999), "Using the SPI to analyze spatial and temporal patterns of drought in Turkey", Drought Network News (1994-2001), Paper 49, available at: http://digitalcommons.unl.edu/droughtnetnews/49

United Nations Office for the Coordination of Humanitarian Affairs (UN OCHA) (2015), El Niño: Snapshot of Impact and Projected Humanitarian Needs.

Vicente-Serrano, S.M., Beguería, S. and López-Moreno, J.I. (2010), "A multi-scalar drought index sensitive to global warming: the standardized rainfall evapotranspiration index – SPEI", *Journal of Climate*, Vol. 23, pp. 1696-1718.

Vicente-Serrano, S.M., Chura, O., López-Moreno, J.I., Azorin-Molina, C., Sánchez-Lorenzo, A., Aguilar, E. and Nieto, J.J. (2015), "Spatio-temporal variability of droughts in Bolivia: 1955-2012", *International Journal of Climatology*, Vol. 35 No. 10, pp. 3024-3040, doi: 10.1002/joc.4190.

Viste, E., Korecha, D. and Sorteberg, A. (2012), "Recent drought and precipitation tendencies in Ethiopia", *Theoretical and Applied Climatology*, Vol. 112 Nos 3/4, pp. 535-551.

Wilhite, D.A. (1993), "Drought assessment, management, and planning: theory and case studies", *Natural Resource Management and Policy Series*, Vol. 2, Kluwer.

Wilhite, D.A. and Glantz, M.H. (1985), "Understanding: the drought phenomenon: the role of definitions", *Water International*, Vol. 10 No. 3, pp. 111-120.

World Bank (2006), *IDA Countries and Exogenous Shocks, IDA Resource Mobilization*, World Bank, Washington, DC.

World Meteorological Organization (2012), *Standardized Precipitation Index User Guide*, in Svoboda, M., Hayes, M. and Wood, D. (Eds), (WMO-No. 1090), Geneva.

Yue, S., Pilon, P. and Cavadias, G. (2002), "Power of the Mann-Kendall and spearman's rho tests for detecting monotonic trends in hydrological series", *Journal of Hydrology*, Vol. 259 Nos 1/4, pp. 254-271.

Yue, S., Pilon, P. and Phinney, B. (2003), "Canadian Streamflow trend detection: impacts of serial and crosscorrelation", *Hydrological Science Journal*, Vol. 48 No. 1, pp. 51-63.

Zargar, A., Sadiq, R., Naser, B. and Khan, F.I. (2011), "A review of drought indices", *Environmental Reviews*, Vol. 19, pp. 333 -349.

Further reading

Andreu, J., *et al.* (2015), "Drought indicators: monitoring, forecasting and early warning at the case study scale", Technical Report No.33.

Edwards, D.C. and McKee, T.B. (1997), "Characteristics of 20th century drought in the united states at multiple time scales", *Atmospheric Science Paper*, Vol. 634, pp. 1-30.

Kumar, N., Murthy, S., Sesha Sai, R. and Roy, S. (2012), "Spatiotemporal analysis of meteorological drought variability in the indian region using standardized precipitation index", *Meteorological Applications*, Vol. 19, pp. 256-264, doi: 10.1002/met.277.

Labedzki, L. (2007), "Estimation of local drought frequency in central Poland using the precipitation index SPI", *Irrigation and Drainage*, Vol. 56 No. 1, pp. 67-77, doi: 10.1002/ird.285.

Lloyd-Hughes, B. and Saunders, M.A. (2002), "A drought climatology for Europe", *International Journal of Climatology*, Vol. 22 No. 13, pp. 1571-1592, doi: 10.1002/joc.846.

Ntale, H.K. and Gan, T.Y. (2003), "Drought indices and their application to east Africa", *International Journal of Climatology*, Vol. 23 No. 11, pp. 1335-1357, doi: 10.1002/joc.93.

Robeson, S.M. (2008), "Applied climatology: drought", *Progress in Physical Geography*, Vol. 32 No. 3, pp. 303-309.

Sternberg, T., Thomas, D. and Middleton, N. (2010), "Short communication drought dynamics on the Mongolian steppe, 1970-2006", *International Journal of Climatology*, Vol. 31 No. 12, pp. 1823-1830, doi: 10.1002/joc.2195.

Tefera, M., Chernet, T. and Harro, W. (1996), "Explanation on geological map of Ethiopia a 1:2,000,000 scale", *Bulletin No 3*, Addis Ababa, pp. 2-64.

Temesgen, D., Hassan, R.M. and Ringler, C. (2008), "Measuring Ethiopian farmers' vulnerability to climate change across regional states", Discussion Paper, International Food Policy Research Institute (IFPRI), Washington, DC.

Tonini, F., Jona Lasinio, G. and Hochmair, H.H. (2012), "Mapping return levels of absolute NDVI variations for the assessment of drought risk in Ethiopia", *International Journal of Applied Earth Observation and Geoinformation*, Vol. 18, pp. 564-572.

Wu, H., Hayes, M.J., Wilhite, D.A. and Svoboda, M.D. (2005), "The effect of the length of record on the standard precipitation index calculation", *International Journal of Climatology*, Vol. 25 No. 4, pp. 505-520, doi: 10.1002/joc.1142.

Corresponding author
Yimer Mohammed can be contacted at: yimermoh2013@gmail.com

10

Local institutional adaptation for sustainable water management under increasing climatic variability and change

Admire Mutsa Nyamwanza

Economic Performance and Development Unit, Human Sciences Research Council, Cape Town, South Africa

Abstract

Purpose – The study aims to explore institutional adaptation for sustainable water resources management at the local level in the context of increasing climate-related challenges in Zimbabwe using the case of a semi-arid area in the mid-Zambezi Valley, north of the country.

Design/methodology/approach – Inspired by the critical institutionalism approach, the study uses qualitative methods (i.e. key informant interviews, semi-structured interviews, community workshops and documentary review) to understand the role of different formal and informal water-related institutions *vis-à-vis* responding to climate-related challenges in the case study area, and how the identified institutions can improve their efforts in the context of national water and environmental policy and regulation frameworks. Thematic analysis was used for data analysis.

Findings – The study found that climatic challenges in the case study area, as in most of rural Africa, have raised the stakes in local water management with respect to regulating access to and balancing competing interests in, and demands for, water. It ultimately argues for the embracing of complexity thinking and flexibility in local water management as well as clear coordination of institutions across scales in the face of increasing climate-related challenges.

Originality/value – The study adds to case studies and evidence-based analyses focused on institutional alternatives for climate adaptation *vis-à-vis* water resources management in water-stressed rural African communities.

Keywords Sustainability, Climate variability, Institutional adaptation, Water management

1. Introduction

Increasing climatic variability and change has caused and is projected to trigger many challenges *vis-à-vis* availability of water, particularly in arid and semi-arid areas. This will, subsequently, place significant management demands on water-related institutions, especially with respect to regulating access for productive and domestic uses and the maintenance of vital ecosystems. As Sadoff and Muller (2009) write, the hard reality is that climate-related challenges will force many trade-offs, for example, around water regulation implementation at the local level. This will require adaptive, flexible and innovative institutions. Institutions are conceptualized in this paper in the mould of North (1990) as formal and informal rules and regulations through which people govern, organise and cooperate to achieve particular purposes. Institutions are critical, as they enable or hinder people's responses to climate-related challenges in various ways. Important functions that water-related institutions undertake, relevant for climate adaptation in local communities, include regulating access to resources and overseeing infrastructural maintenance, skills development and capacity building, influencing other decision-makers and institutions and information gathering and dissemination.

It is, therefore, critical to focus attention on the nature and structure of institutions governing access to and use of key resources such as water in the context of increasing climate-related challenges. This is important if adaptation processes and interventions are to succeed and help particularly the most vulnerable individuals, households and groups at the local level. It is in this context that this study explored institutions and options for institutional adaptation at the local level *vis-à-vis* sustainable water management in rural Zimbabwe using the case of Mbire District in the mid-Zambezi Valley area of Mashonaland Central Province. The term "institutional adaptation" is taken to mean modifications and alterations within institutions or its components to adjust to changes in the wider physical and social environment (after Adger, 2000). On the other hand, sustainable water management in this study alludes to:

> Innovative investments for capacity extension in water access and use, ensuring efficient and equitable operation of existing water systems for optimal use, overseeing the continued maintenance and rehabilitation of water systems and modification in processes and demands towards reconciling social justice and environmental integrity concerns (Nyamwanza and Kujinga, 2017, pp. 695-696).

Sustainable water management under increasing climate-related challenges requires responsive and adaptive institutions, as water management primarily takes place within local and national institutional arrangements (Nyamwanza and Kujinga, 2017). The study essentially adds to evidence-based analyses on local institutional alternatives for climate adaptation *vis-à-vis* water management in water-stressed rural African communities.

The overriding question of interest was, what should local water-related institutions do to evolve and become more capable and effective in assisting local communities to deal with increasing climate-related challenges? Pursuant to this main question, the following four research questions were explored:

RQ1.   What are the formal and informal institutional arrangements that govern access to and use of water resources at the local level in the case study area?

RQ2.   How are the institutions structured and what are the results of cross-scale institutional linkages and collaborations (if any), particularly with respect to assisting community members to effectively respond to the impacts of climate-related challenges?

RQ3. How have the different institutions facilitated particular types of adaptation strategies in the area and what have been the levels of success of their efforts over the years?

RQ4. What alternatives are there *vis-à-vis* sustainable water management and improving the institutions' efforts in adaptation processes in the area in the context of current national water and environmental policies and regulations?

The study was influenced by the critical institutionalism (CI) approach. CI is a body of thought that seeks to explore how institutions dynamically mediate relationships between people, natural resources and society (Cleaver and de Koning, 2015). It posits that institutions are dynamic and that they must not necessarily be designed and/or perpetuated for single specific purposes, but should be flexible and open enough towards integration with other working arrangement as conditions evolve (Hall *et al.*, 2013; Cleaver and de Koning, 2015). To that end, the interplay of institutions is given particular attention in CI, and the idea of "institutional bricolage" is given prominence within the approach. Cleaver, 2012, p. 45) defines institutional bricolage as:

> A process in which people consciously and non-consciously draw on existing social formulae (style of thinking, models of cause and effect, social norms and sanctioned social roles and relationships) to patch or piece together institutions in response to changing situations.

Institutional bricolage therefore involves mixing old and new arrangements to come up with more practical and useful institutional frameworks (Cleaver and de Koning, 2015). In the context of institutional adaptation for sustainable water management under increasing climate-related challenges, a promising direction for CI, as applied to this study, would be to draw on different available institutional actors, materials and resources, regardless of their original purposes, to come up with innovative and responsive institutional arrangements. Adapted configuration of rules, practices, norms and relationships are then attributed meaning and authority (Cleaver and de Koning, 2015).

The paper is divided into six sections. Following this introduction section is an exploration of the national water, environmental and climate context within which the study was undertaken, in Section 2. This includes a presentation of evidence of climate change in Zimbabwe in general and implications on water resources, and a discussion of current national policies and regulatory frameworks informing water, the environment and climate change. Section 3 is the methodology section, which includes contextual information on the case study area and methods used for data collection and analysis. Section 4 presents results from the fieldwork exercise and documentary review towards addressing the research questions. Section 5 then discusses alternatives for improving local water-related institutions' efforts in adaptation processes in the case study area. Lastly is the conclusion section.

2. The national water, environmental and climate change context
2.1 Evidence of increasing climate challenges and implications on water resources in Zimbabwe

According to Zimbabwe's National Climate Change Response Strategy document, the country's continental interior location means that it is predicted to warm more rapidly in the future than the global average (Government of Zimbabwe, 2015). Already, the Zimbabwe Meteorological Services has reported that daily minimum temperatures have risen by approximately 2.6°C over the past century, while daily maximum temperatures have risen by 2°C during the same period, with anecdotal evidence also pointing to wet seasons progressively starting later (Brown *et al.*, 2012; Davis and Hirji, 2014). Chagutah (2010) notes that by the end of the twentieth century, Zimbabwe had become warmer and drier than it was at the beginning, with annual mean temperatures having increased by about 0.4°C since 1900 and rainfall having declined by nearly 5 per cent across the country. Citing studies undertaken by the University of Zimbabwe's Department of Geography, Brown *et al.* (2012) note that there is a predicted warming of around 2°C by 2080 in Zimbabwe, with annual rainfall averages predicted to be between 5 per cent and 18 per cent less than the 1961-1990 average by the same year (Government of Zimbabwe, 2015). Furthermore, whilst Zimbabwe has experienced periodic floods and, on average, one to three droughts every five years largely due to changes in the phase of the El Nino Southern Oscillation (ENSO) phenomenon and periodic sea surface temperature changes, these are expected to increase in frequency and intensity due to global climate variability and change (Chagutah, 2010).

Owing to the noted climate dynamics, there is projected to be a severe strain on the country's surface and groundwater systems, and this will have a heavy effect on the demand for, and supply and quality of water in local communities. Already, there has been a decline in stream flow over the past 20 years in the country, although it is not clear that this decline is because of climatic changes (Government of Zimbabwe, 2015). What is more certain however is that due to increased warming, evaporation is predicted to increase by between 4 to 25 per cent in the river basins, and run-off projected to decline by up to 40 per cent. The Zambezi basin (within which the case study area is located), is projected to be the worst affected (Brown *et al.*, 2012). Consequently, both rainfed and irrigated agriculture will face reduced yields.

It is also estimated that existing dams constructed for water supply and irrigation may become less reliable, and there is likely to be increased reliance on groundwater since, unlike surface water, it is shielded from evaporative losses (Government of Zimbabwe, 2015). Yet surface water contributes over 90 per cent of the country's water supply (Mtisi and Prowse, 2012), whilst the potential to fully utilise groundwater in Zimbabwe has not yet been realised mainly due to the unaffordability of the required technology (Brown *et al.*, 2012). It is important at this point to note that Zimbabwe had developed an extensive water storage capacity prior to and after independence. This, however, went on to severely deteriorate in many areas post 2000 due to the serious economic and financial crises that engulfed the country between 2000 and 2009. For instance, dam construction and inspection, irrigation scheme establishment and maintenance and other major water projects either stalled or became redundant (Davis and Hirji, 2014).

2.2 Environmental management in Zimbabwe

Climate and water are part of the wider environmental system. It was therefore important to understand the general environmental policy framework in Zimbabwe so as to obtain a clear picture with respect to the national environmental management context within which local water-related institutions operate. Environmental management in Zimbabwe is primarily regulated by the Environmental Management Act of 2002. This Act provides the overall framework for the sustainable management of natural resources, including water, and protection of the environment. The Environmental Management Act is administered by the Environmental Management Agency (EMA), a statutory body in the Ministry of Environment, Climate and Water. Among its core principles, the Environmental Management Act states that all elements of the environment, e.g. water, climate, vegetation, air and soils are linked and interrelated and, therefore, environmental management must be integrated and the best practicable environmental management options pursued. It also states that environmental management must place people and their needs at the forefront of its concerns. Furthermore, it notes that sensitive, vulnerable and highly dynamic or stressed ecosystems require specific attention in management or planning procedures, especially when they are subject to significant human resource usage and development pressure (Section 4.2, Environmental Management Act). EMA has officials at both provincial and district levels who have a duty to assist local authorities in formulating by-laws which dovetail with national environmental policies as well as ensuring compliance to these laws.

2.3 National climate change policy framework

Zimbabwe has a fairly sound climate change policy framework. It has actively participated in international negotiations on climate change, having signed and ratified the United Nations Framework Convention on Climate Change (UNFCCC) in 1992 and acceded to the Kyoto Protocol in 2009. The country produced its Initial National Communication to the UNFCCC in 1998 and the Second National Communication in 2013. A National Climate Change Response Strategy (NCCRS), which provides a framework for a comprehensive and strategic approach on climate adaptation, mitigation, technology, financing, public education and awareness, was finalized in 2015. The NCCRS aims at bringing about an integrated response strategy across all sectors.

The main entity at the forefront of formulating and leading the implementation of climate change policy in Zimbabwe is the Department of Climate Change Management housed in the Ministry of Environment, Water and Climate. Its responsibilities include engaging with the UNFCCC and producing National Communications and coordinating specific national climate change projects between and across various ministries and organizations. Two Acts also implicitly deal with climate change issues in the country and these are:

(1) The Civil Protection Act of 1989 administered by the Ministry of Local Government, Public Works and National Housing. The Act provides for disaster management activities, thus including solutions to extreme weather events due to climate change and variability.

(2) The Meteorological Services Act of 1990 establishing the administration of meteorological services and providing for the carrying out of meteorological research and investigation, including the issuing of weather forecasts and climate projections.

2.4 National water policy and regulation frameworks

National water policy and regulation frameworks in Zimbabwe are primarily governed by two pieces of legislation – the Water Act and the Zimbabwe National Authority (ZINWA) Act – both promulgated in 1998. The country also has a National Water Policy document launched in 2012, which states its overall objective as assisting in the achievement of sustainable utilization of water resources (Government of Zimbabwe, 2012). The Water Act vests ownership of all water in the President rather than private land owners. It also provided for the establishment of catchment and sub-catchment areas (overseen by catchment councils [CCs] and sub-catchment councils [SCCs]). These catchment areas were drawn according to hydrological boundaries and they form the basis for water planning, issuing of water-use permits, establishing the rights and responsibilities of water users and assigning responsibility for dam safety (Davis and Hirji, 2014). To that end, seven catchment areas were established in the country.

The ZINWA Act established the Zimbabwe National Water Authority (ZINWA) – a parastatal whose functions vary depending on the level at which it is operating. At the national level, ZINWA advises the Minister responsible for water on formulation of national policies, water resources development and management and water pricing. At the local level, its roles include ensuring that CCs and SCCs discharge their functions in accordance with the Water Act. Zimbabwe's national water policy and regulation frameworks are ultimately based on the underlying concept of "integrated water resources management" which emphasizes four main principles namely: management of water on the basis of hydrological boundaries, decentralization of water management, stakeholder participation and representation in water management processes and treatment of water as an economic good (Mtisi, 2011). Apart from the Water Act and the ZINWA Act, three other Acts also inform water resources management in (rural) Zimbabwe. These include:

(1) The discussed Environmental Management (EMA) Act of 2002 which provides the overall framework for the sustainable management of natural resources and protection of the environment (see Section 2.2).

(2) The Rural District Councils (RDC) Act of 1996 which empowers rural councils to formulate by-laws governing, among other aspects, natural resources within their jurisdiction.

(3) The Public Health Act of 2002 which, among other aspects, regulates issues around water quality monitoring, safe water supply and household sanitation.

There is also the Ministry of Transport, Communication and Infrastructural Development (MTCID) which maintains a small unit through the District Development Fund (DDF). The unit is responsible for providing technical guidance and expertise to RDCs in planning and supervising rural water, sanitation and hygiene and borehole drilling, as well as pump maintenance and rehabilitation (Government of Zimbabwe, 2012). The National Water Policy document of 2012 recognizes climate change and calls for the establishment of specific provisions for understanding the extent of the threat and specific actions for managing potential impacts (Davis and Hirji, 2014).

3. Methodology
3.1 Case study area
The area selected for case study (Mbire District) has 17 wards/local administrative geographical boundaries. This is a semi-arid remote area bordered by Zambia, Mozambique, Mashonaland West Province and other Mashonaland Central Districts of Guruve and Muzarabani (Figure 1). Mbire District is characterized by a dry, tropical climate with temperatures which can range up to 40°C in summer and low rainfalls averaging 450-650 mm annually. From documentary analysis and interviews with key informants and community members, main weather and climate dynamics in the area have primarily been characterized by increasingly high air temperatures, increasing erratic rainfall patterns and increasing frequency in the occurrence of extreme weather events, particularly droughts and floods. This case study was selected based on the fact that the impacts of climate challenges are likely to be more severe in social and ecological contexts of arid and semi-arid regions. These are areas where livelihoods are often already stressed and additional adverse biophysical and social changes can be overwhelming (Agrawal, 2008). This makes the identification and understanding of viable adaptation solutions and alternatives in such contexts urgent.

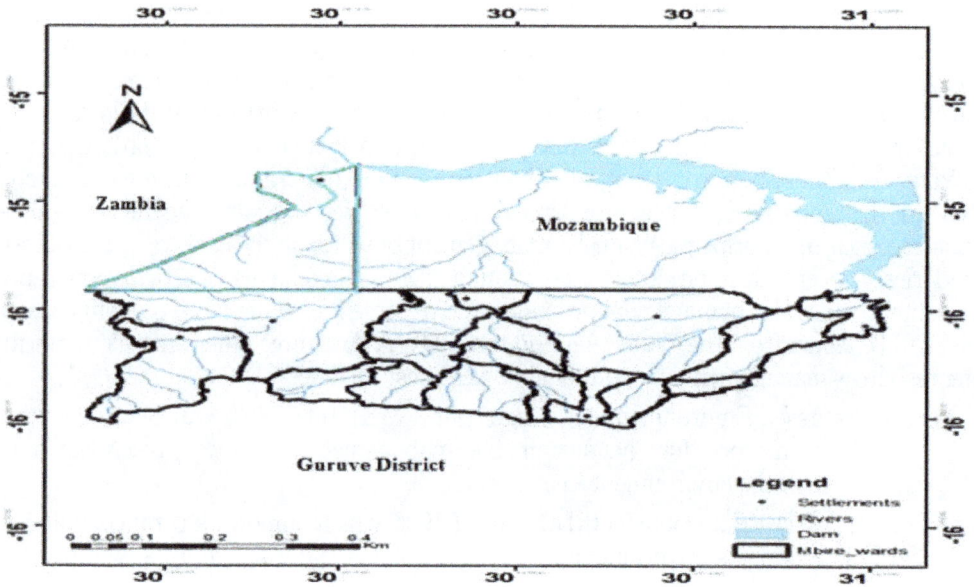

Figure 1. Mbire District, Zimbabwe

3.2 Methods

Fieldwork for this study was undertaken in the period September-October 2015. The study used semi-structured interviews with community members, key informant interviews (KIIs), community workshops and documentary analysis. KIIs were conducted, firstly, at the national level, with officials from the Zimbabwe National Water Authority, Environmental Management Agency and the Department of Climate Change Management. Secondly, KIIs were conducted at the local level in Mbire District with selected officials involved in identified water-related institutions in the area (i.e. Mbire Rural District Council, Environmental Management Agency, Traditional Authority and Borehole Water Committees). Questions for national-level key informants revolved around the roles of their institutions in national climate change and/or water policy formulation, their views on the alignment of national water policies and regulations with climate dynamics in water-stressed rural communities in Zimbabwe and their roles and activities (if any) in the specific case study area of Mbire District. Questions for local-level key informants revolved around the role of their institutions in water management in the area, any changes/alterations and challenges and opportunities *vis-à-vis* their water management processes and practices owing to climate challenges in the area, and collaborations with other institutions.

In total, 34 semi-structured interviews and 3 community workshops were conducted with randomly selected adult respondents in the area. Questions in semi-structured interviews and community workshops included those around water resources and sources of livelihoods in the area, institutions involved in water management in the area (including the effectiveness of these institutions in the context of increasing climate challenges), adaptation strategies as facilitated and/or enabled by identified water-related institutions and respondents' views on how these institutions may improve their efforts. A documentary review of relevant published and unpublished materials was also used in exploring issues around the national water, environmental and climate change contexts in Zimbabwe. All data were analysed using thematic analysis method.

4. Results

4.1 Sources of livelihoods, water sources and their uses in Mbire District

4.1.1 Sources of livelihoods in Mbire District. Understanding sources of livelihoods was important in establishing key productive uses of water in the area. Sources of livelihoods in the area revolve around agricultural activities – encompassing crop and livestock production as well as local casual agricultural work (in other villagers' fields for payment in cash or in kind). People are also involved in non-agricultural activities which include fishing (for both household consumption and commercial purposes), trading in goods (in small shops and flea markets) and remittances for a few households. Crop production involves two systems: the upland and the riverbank crop production systems. The upland crop production takes place in upland fields or *minda yekunze* in the Shona vernacular, averaging 2 to 5 ha per household. The upland fields consist of shallow sandy clay soils, and are used for dryland crop production involving mainly cotton, sorghum and groundnut farming. Riverbank crop production takes place in fields on the banks of major rivers, referred to as *minda yekugova* in the area, and averaging 0.5 to 2 ha per household.

Riverbank fields form the most arable lands in the area as they are characterised by rich alluvial fertile soils which store residual moisture from the rainy season into the dry season, thereby enabling villagers to conduct farming there the whole year round. Kinship and lineage are important factors in the ownership of riverbank fields as the fields are inherited and passed on from generation to generation. The majority owners of riverbank fields are autochthonous residents, with others gaining usufruct rights as mainly based on close neighbourhood, friendship or kinship. Major crops grown in riverbank fields include maize, green vegetables and tomatoes. All households interviewed were involved in both upland and riverbank crop production during the time of the research. Livestock production in the area includes cattle, goat, sheep, pig and poultry rearing, with cattle, goats and poultry forming the larger number of animals.

4.1.2 Sources of water in Mbire. The main sources of water in the area include rivers, seasonal streams, boreholes, protected and unprotected wells. Table I shows the main water sources for different uses and the importance of the different sources as ranked by respondents in semi-structured interviews according to reliability of the water source and frequency of use throughout the year.

Table I. Water sources and their uses in Mbire District

Water use	Sources (ranked according to importance)
Upland crop production	Rainfall
	Seasonal streams
Riverbank crop production	Rivers
	Rainfall
Livestock watering	Rivers
	Seasonal streams
	Boreholes (during dry seasons)
	Unprotected wells (in dry seasons)
Domestic use	Boreholes
	Protected and unprotected wells
	Seasonal streams
	Rivers

4.2 Climate-related challenges in Mbire District
This section discusses climate-related challenges in Mbire District over the years and their impacts *vis-à-vis* water and livelihoods. Climate challenges and their impacts were understood through exploring unfavourable weather and climate dynamics in the area over the years, and how these dynamics have had an effect on general water availability for both agricultural and domestic purposes. Evidence from both primary fieldwork exercises and secondary data shows that the most serious climate challenges have manifested through increasing intra-seasonal dry spells, increasing drought cycles and floods.

4.2.1 Intra-seasonal dry spells. Respondents in community workshops and semi-structured interviews pointed out that the frequency of dry spells during cropping seasons has increased over the years. These dry spells were noted as mainly occurring early in the season, particularly between October and December – coinciding with the germination stage when crops are most sensitive to moisture stress. A season was therefore said to be able to receive normal (or even increased) rainfall amounts; however, these intra-seasonal dry spells, coming at a time when crops are at a critical stage of growth, have many a time led to reduced quantity and quality of crop yields. Respondents also noted shifts in the onset of rains as well as seasons becoming shorter as other climate-related developments compounding the problem of intra-seasonal dry spells.

4.2.2 Increased drought cycles. Local-level key informants and community members during interviews stated that droughts were increasing in frequency in recent years. It was noted that droughts were now occurring on average every three years (unlike the five-year average which was usually experienced in the 1980s and 1990s). These droughts have led to total crop failure (especially in upland fields), drying up of rivers and boreholes and livestock deaths. A historical timeline exercise identifying serious drought years which have occurred in the area since 1980 carried out with community members during community workshops showed the following years as having been affected: 1982-1983, 1991-1992, 1994-1995, 2001-2002, 2003-2004, 2006-2007, 2008-2009 and 2012-2013. The trend clearly showed an increasing frequency in the occurrence of droughts. When asked on the general availability of water for both agricultural and domestic purposes over the preceding five years prior to the fieldwork, slightly over 60 per cent of respondents in semi-structured interviews noted that water for both agricultural and domestic purposes was generally becoming scarcer as opposed to being "enough" or "abundant" (Figure 2). The main reasons proffered for this were the drying up of streams, boreholes and wells owing to increasing drought cycles.

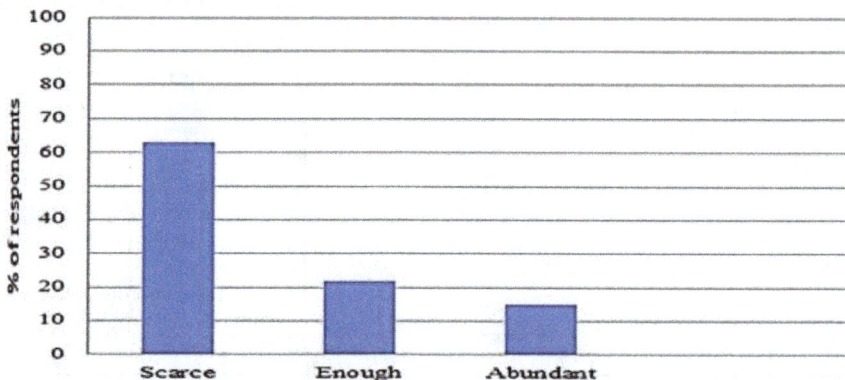

Figure 2. Community members' perceptions on the general availability of water in Mbire District

4.2.3 Floods. Floods in Mbire have mainly occurred due to two factors:

(1)   excessive local rains occurring mostly around January/February; and
(2)   water release from Kariba Dam upstream and backflow from Cabora Bassa Dam in Mozambique further downstream.

The second factor arises from the fact that Mbire District is located downstream of Kariba Dam and upstream of Cabora Bassa Dam. From December to February, Kariba Dam levels usually rise higher than normal and water is released to avoid dam failure. This results in substantial increase in discharge into the Zambezi River and backflow pressure from Cabora Bassa Dam. Subsequently, major rivers in the district (i.e. Manyame, Mwanzamutanda, Kadzi and Musengezi) will not be able to discharge into the Zambezi River which results in water accumulating at the confluence of these rivers to the Zambezi River, leading to flooding in the Mbire District area. A discussion with the RDC CEO and with community members across different wards revealed that flood frequency has increased in the district in the past two decades, with the most recent being experienced in 2014.

4.3 Water-related institutions in Mbire District: structure and roles
This sub-section discusses the various water-related formal and informal institutional arrangements involved in water management in Mbire as identified by respondents – including their current roles and evolving structures. The section responds to research objectives one and two and sets the building blocks for analysing how the identified institutions may adapt or be positively transformed towards assisting people in the area to effectively deal with climate challenges discussed in Section 4.2. It is noteworthy at this point that whilst, as discussed in Section 2, CCs and SCCs, as well as ZINWA, are (supposed to be) the key institutions with respect to water management as established by the Water Act and the ZINWA Act, these institutions are hardly known and/or acknowledged as active in water management in Mbire. Figure 3 shows the percentages of people (among the 34 interviewed across the district), who had and had no knowledge of (the existence of) CCs, SCCs and ZINWA in the area. Over 90 per cent of respondents had no knowledge of the responsibilities, operations and activities of CCs and SCCs in the area. On the other hand, whilst a significant number of respondents acknowledged that they had heard of ZINWA (especially in the media in relation to its activities in towns and cities), they were not aware of its responsibilities as a water authority in the area and had never interacted with an official from ZINWA.

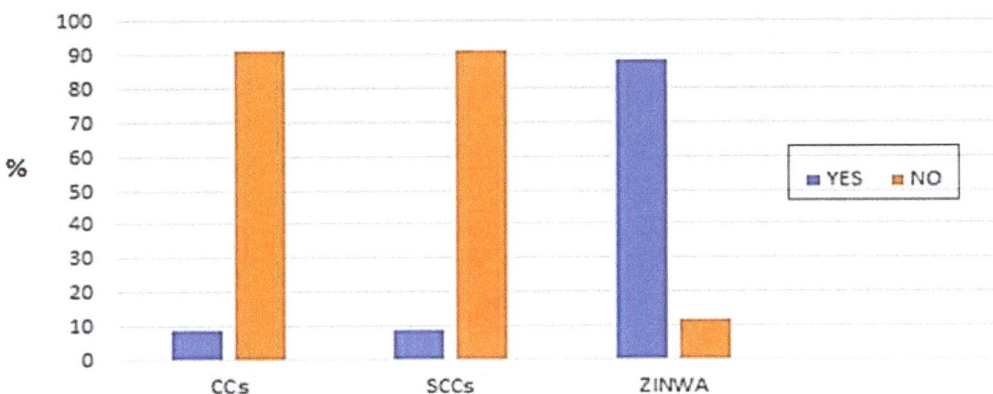

Figure 3. Knowledge of CCs, SCCs and ZINWA

An agreed-upon fact therefore, even among interviewed key informants, was that though Mbire District does lie within Manyame catchment area, and within Lower Manyame and Angwa-Rukomechi sub-catchment areas, CCs, SCCs and ZINWA are not active with respect to water management in the area. This resonates with observations from other rural communities in the country (Kujinga, 2004; Twikirize, 2005). A ZINWA official in the capital Harare noted during an interview that communal areas such as Mbire have not had sustained water use for commercial purposes and have not had a stake in commercial water which is why these institutions are not active in these areas. He also noted that the lack of financial resources, especially between 2000 and 2009, crippled the operations of these institutions such that they failed to actively penetrate and establish functional structures covering such marginal areas as Mbire District. These institutions are therefore not included in discussions in this paper. Discussions in this section cover the following water-related institutions as identified in the area: The Rural District Council (RDC), the Environmental Management Authority (EMA), traditional authorities (TAs), the Ministry of Transport, Communication and Infrastructural Development's District Development Fund unit (MTCID's DDF) and Borehole Water Committees (BWC). Table II summarizes the various responsibilities of these water authorities *vis-à-vis* specific water resources in the area.

4.3.1 Rural District Council. The RDC was the institution mentioned by the highest number (over 80 per cent) of respondents when asked to identify water-related institutions in the area. According to the Mbire RDC Chief Executive Officer, the RDC's roles in water management in the area range from instituting by-laws regulating such water resources as rivers and streams, settling disputes and conflicts that may arise among water users, e.g. between upstream and downstream water users, to training other water-related institutions like borehole water committees in water-point management. He noted that the RDC was also the responsible authority overseeing the spearheading of major water investments planned for the area, particularly the construction and resuscitation of dams and irrigation schemes, as well coordinating

Table II. Summary of water-related institutions and their responsibilities in Mbire District

Water authority	Specific water resource (s) managed	Responsibilities
RDC	Rivers and streams	instituting and enforcing by-laws, solving disputes and conflicts that may arise between and among users
	Boreholes	training borehole committees on water-point management
EMA	Rivers and streams	monitoring compliance to and enforcing national regulations and local by-laws
TAs	Rivers and streams	instituting and enforcing regulations on use, solving disputes and conflicts that may arise between and among users
MTCID's DDF unit	Boreholes	drilling and ensuring maintenance and rehabilitation
BWCs	Boreholes	instituting and enforcing regulations on use, ensuring maintenance and rehabilitation, solving disputes and conflicts that may arise between and among users, monitoring and guarding against water theft

responses to water-related extreme weather events such as droughts and floods. With these diverse roles, the RDC can therefore be said to be the main regulating authority of water resources in Mbire District. The RDC also collaborates with other formal and informal water-related institutions in the area *vis-à-vis* coming up with and enforcing water-related regulations in the area.

4.3.2 Environmental Management Authority. The EMA, as discussed earlier, derives its authority from the EMA Act (2002), and it was identified by respondents as also very visible in water-resources management in the area. According to the EMA Mbire District Officer, EMA's primary role, as derived from the EMA Act, relates to overseeing the general protection of the environment. This includes prevention of water pollution and proper utilization of commonly held resources as rivers and streams. The high visibility of EMA in Mbire District has mainly been due to the widespread engagement in riverbank farming in the area and contestations between villagers and authorities around this practice (which will be discussed in detail in the following sections). EMA officers work hand-in-glove with the RDC in monitoring compliance of and enforcing by-laws and other regulations that fall within their mandate.

4.3.3 Traditional authorities. TAs identified as active in the management of water resources in the area include chiefs, headmen, village-heads and spirit mediums. These were noted as having quite a huge influence on the utilization of rivers and streams in the area – particularly with respect to instituting and enforcing regulations of use, as well as settling disputes and conflicts that may arise in the process of utilizing the resources. Chiefs are the heads in the line of traditional authority followed by headmen and lastly village heads. These traditional leaders work hand-in-glove with spirit mediums, known in the area as *homwe dzavanasekuru* or *masvikiro.* Spirit mediums are responsible primarily for communicating and relaying messages from the departed royal ancestors who are believed to be the real owners of the land or *varidzi venzvimbo* in vernacular. The departed royal ancestors are believed to be the providers of rain and fertility of the land and responsible for the general protection of people against various social ills. Because of that, spirit mediums are consulted in times of intense community vulnerabilities such as droughts and floods thus their views on access to and use of commonly held natural resources as rivers and streams are hugely respected.

4.3.4 The Ministry of Transport, Communication and Infrastructural Development's District Development Fund unit. As already noted earlier, the DDF maintains a small unit within the MTCID responsible for the provision of technical guidance and expertise to RDCs in planning and supervising water sanitation and hygiene, borehole drilling, pump maintenance and rehabilitation. In Mbire District, the DDF unit was identified as having drilled close to 70 per cent of boreholes in the area and following up on the maintenance of these boreholes especially in the first few years of the setting up of the boreholes.

4.3.5 Borehole Water Committees. All functioning boreholes in the district are run by specific committees. BWCs are normally elected annually, and are chosen from among villagers consistently using a particular borehole within a "reasonable" radius as subjectively judged by borehole users themselves. BWCs therefore may or may not overlap village boundaries depending on where the borehole is positioned. BWCs typically consist of a chairperson, a vice chair, a treasurer and two committee members. Their major responsibility is to ensure the proper usage and maintenance of their particular borehole. When a borehole breaks down, they mobilise for its repair (e.g. collecting money from users to fund repairs, and looking for people to repair the borehole). They are also responsible for the periodic collection of funds for the purchasing of lubrication oil needed to consistently apply on the boreholes for their effective and efficient functioning. Furthermore, they are involved in settling disputes and conflicts that may arise among borehole users. Whilst it is every user's responsibility to monitor that water from their borehole is not being stolen, the BC is the ultimate authority in ensuring that water from the borehole is not being stolen by unauthorised users.

4.4 Water-related institutional efforts in responses to climate challenges in Mbire
This section discusses how the various water-related institutions identified in Section 4.3 have initiated and/or facilitated particular types of responses and/or adaptation strategies *vis-à-vis* climate challenges discussed in Section 4.2. The discussion starts with responses and/or strategies as initiated and/or facilitated by formal institutions before discussing those undertaken by TAs. Levels of success of the institutions' endeavours as assessed by officials involved in the various institutions and by community members are also discussed.

4.4.1 Efforts by the RDC. The RDC has initiated and facilitated various efforts towards assisting communities in Mbire District to adequately respond to all the climate challenges discussed in Section 4.2. In relation to dealing with dry-spells and droughts in the area, firstly, the RDC has instituted by-laws over the years to ensure responsible use and preservation of water resources. Water-related by-laws instituted include:

- the prohibition of streambank cultivation;
- the prohibition of illegal gold panning activities; and
- the prohibition of farming in wetlands.

The RDC has also facilitated and supervised most government and non-governmental drought relief programmes in the area over the years. In addition, the RDC has also trained most borehole water committees across the district in water point management, and appraising community members on the water, sanitation and hygiene (WASH) concept. According the RDC CEO, the RDC has also pushed strongly for the rehabilitation and/or revival of major water infrastructure in the area such as dams and irrigation schemes most of which have become dysfunctional and obsolete over the years – although this has met little success.

In as far as floods are concerned, the RDC is the local coordinating authority on the standing committee which deals with floods in the area. This committee includes representatives from the Provincial Civil Protection and Planning Committee (PCPPC), local health officials and community based organizations (CBOs) and various NGOs. The RDC is therefore actively involved in the provision of early warning to communities, community preparedness, evacuation and immediate and long-term relief assistance. Interviews with the district administrators and selected councillors however revealed that the flood response team is not adequately capacitated particularly with respect to the provision of such critical resources as vehicles, training (e.g. in conducting flood risk assessments) and financing (e.g. to enable the purchasing of reliable communication equipment for communities at high risk in the area).

4.4.2 Efforts by EMA. As already discussed in Section 2, the main mandate of EMA revolves around the general protection of the environment. In Mbire District, its work in as far as water management is concerned has revolved around:

- educating residents through awareness campaigns and community workshops on the advantages and benefits of protecting water resources in a drought and flood-prone area; and
- enforcing environment related national and local by-laws with the assistance of officers from the Zimbabwe Republic Police.

Interviews with villagers and discussions in community workshops, however, revealed that there is now clear animosity between EMA officials and local communities due to what residents described as the institution's "zealous" enforcement of the contentious national streambank cultivation regulation (which prohibits cultivation "within 30 m of the naturally defined banks of a public stream") (Government of Zimbabwe, 1991). The head of the local EMA office admitted as much when he noted that their officers had, many times, even been threatened with mysterious deaths if they dared continue their forceful push to stop riverbank farming during their routine visits to monitor the extend of siltation being caused by riverbank farming activities across different wards.

This (animosity) is in the context of the high importance attached to riverbank fields in the area in as far as all-year-round maize and vegetable farming is concerned (as discussed in Section 4.1). Whilst the EMA officer interviewed noted that they were doing fairly well *vis-à-vis* educating and conscientising people on the need to find other innovative and sustainable response strategies to droughts and dry-spells without resorting to extensive riverbank farming and illegal gold panning, he articulated two main challenges in the course of carrying out their duties. These included:

- the reluctance of other institutional players to forcefully push for compliance to natural resource by-laws in the area – apparently because of the fear of the previously mentioned death threats; and
- resource shortages.

On resource shortages, he noted that, for example, the local office only had four on-the-ground officers, and these officers are expected to monitor natural resource protection across all the 17 wards in the district. He also noted that they only had one aging motorbike for transport, and they have had to rely mostly on the RDC for transport to carry out their work in the area.

4.4.3 Efforts by the MTCID's DDF unit. As noted in Section 4.3, the DDF unit has been instrumental in drilling and maintaining boreholes across all the 17 wards in Mbire District over the years, with over 70 per cent of the boreholes in the district having been sunk by the unit. Boreholes in drought-prone areas as Mbire are critical as important sources of water for both people and livestock. The activities of this unit in Mbire have, however, stalled owing to lack of resources. DDF units were 80 per cent dependent on donor funding (cf. Kwangare *et al.*, 2014), and when funding from most Western countries was suspended in the early 2000s primarily over the chaotic and violent "fast-track" land reform programme in Zimbabwe, their activities were heavily affected. Most boreholes have thus become dysfunctional because of lack of proper maintenance, with the situation being compounded by periodic floods in the area – where the few functional boreholes are at risk of being flooded and contaminated each time floods occur. The borehole situation in the district has, therefore, generally become dire. In virtually all wards for example, with the exception of Ward 1 only, the maximum number of people per borehole far supersedes the recommended figure (of 250 persons or 25 families per borehole according to the Government's Water and Sanitation Master Plan). According to the RDC Projects Officer, the worst affected ward (Ward 11/Masoka ward) has over 2,000 persons per functioning borehole.

4.4.4 Efforts by BWCs. According to villagers in community workshops, BWCs have been quite helpful in keeping most boreholes running and functional, especially in the background of receding DDF unit activities as discussed, coupled with massive population increases and the subsequent pressure on boreholes. The various borehole rules set by these BCs have, for example, ensured responsible use, regular maintenance and timely repair of boreholes. This has led to some level of water security, particularly for domestic purposes over the years in the area. Respondents however noted that owing to contamination of most boreholes during periodic floods and low rainfalls and recurring droughts (and the subsequent lowering of the water table in most places) in the area, most boreholes had become dysfunctional, leaving most committees redundant and leading to their folding up.

4.4.5 Efforts by TAs. Efforts by TAs (i.e. chiefs, headmen, village heads and spirit mediums) towards initiating and/or facilitating responses to climate challenges *vis-à-vis* water management in Mbire District have revolved around:

- preservation of such community water resources as rivers and streams; and
- raising community concerns *vis-à-vis* water resources and livelihoods with other water-related institutions.

In relation to the first point, the traditional leadership (together with other water-related institutions) has instituted, supported and helped enforce certain regulations particularly related to the responsible use and preservation of rivers and streams. This includes the previously mentioned regulation prohibiting illegal gold panning in rivers, proper authorization for sinking unprotected wells in the area, as well as warning and punishing culprits caught breaking rules within their areas of jurisdiction. On the second point, the traditional leadership has been very instrumental in convincing other water institutional players, particularly the RDC, on the significance of local livelihoods dynamics, particularly around riverbank farming, especially in the context of increasing droughts and dry-spells in the area. This has led to the modification of the previously mentioned national streambank cultivation regulation – where the RDC agreed that people can be allowed to continue farming in the riverbanks unhindered provided they do not use ox-drawn ploughs or any other farming implement besides the hoe. Although this agreement is still to get a full buy-in from other water-related institutions in the area, particularly EMA, it has helped people to respond to the effects of droughts and dry-spells in the area.

5. Discussion and analysis

5.1 Adapting local institutions towards sustainable water management under increasing climatic challenges in Mbire District

From discussions in Section 4, it is clear that increasing climate challenges in Mbire District have resulted in the general scarcity of water in the area for all uses and increased floods and drought cycles. This has led to the redundancy of most boreholes and a heavily compromised upland farming system, resulting in increased reliance on riverbank farming. Climate challenges in Mbire have consequently raised the stakes in local water management *vis-à-vis* balancing competing interests in, and demands for, water. This section explores options and alternatives for improving the discussed water-related institutions' efforts in helping communities in the area to effectively respond to (the effects of) increasing climate challenges. Three main categories of efforts can be identified from the various initiatives discussed. In their efforts, local water-related institutions have been involved in one or all of the following initiatives:

- ensuring the preservation and responsible use of water resources through water-point regulation and by-law formulation and enforcement;
- coordinating local flood response and drought-relief programs; and
- continuously engaging central government on the revival of key water infrastructure in the area (e.g. collapsed irrigation schemes, dysfunctional boreholes and dams).

Two factors are raised as options towards adapting local institutions *vis-à-vis* improving their efforts towards sustainable water management. These (factors) form the discussions in the following subsections.

5.1.1 Embracing complexity thinking and flexibility in local water management. Complexity thinking in natural resources management relates to accepting unpredictability and uncertainty as part of dealing with people and situations, as well as acknowledging a multitude of perspectives in decision-making (Simonsen *et al.,* 2014). Complexity thinking advocates for flexible process-based learning as an essential part of decision-making, recognition of context-specific dynamics and taking up an experimental approach to resource management (cf. Nyamwanza and Kujinga, 2017). It therefore negates from, for example, new institutionalism thinking of predictability and use of routine processes and principles, towards an active recognition of unique contextual political, natural and socio-economic processes in all aspects of resource management. As the CI school propounds, institutional actors and structures should have the latitude to fluidly adjust in addressing new challenges and situations. In the case of Mbire, local water-related institutions have to take up complexity thinking and flexibility in both the formulation of water-point regulations and by-laws and their enforcement. A case in point is that of riverbank farming. Mbire is a dry area, and with climate challenges discussed in Section 4.2 – particularly increasing drought cycles and intra-seasonal dry-spells – riverbank farming has become extremely important in sustaining livelihoods. It is interesting to note that whilst the RDC and the traditional leadership agreed on modifying the national regulation on streambank cultivation in line with local dynamics, residents have continued to face pressure from EMA to stop farming activities in the riverbanks completely. Whilst officials from both EMA and the RDC noted during key informant interviews that they do work hand-in-glove, it is clear that there has not been convergence in dealing with the riverbank farming issue in the area – which feeds into the next point discussed in the following subsection.

5.1.2 Clear coordination of institutions in water management. As shown in earlier discussions, the water institutional environment in Mbire is characterized by a plurality of organizations and actors, as well as multiple rules and procedures some of which are overlapping. An example is that of the RDC, EMA and the traditional leadership who all regulate the use of, and access to rivers and streams in one way or the other. As already discussed in the previous subsection, one consequence of this multiplicity of institutions in water management has been confusion in applying and enforcing certain regulations. In the context of dynamics resulting from increasing climate-related challenges however (such as a surge in riverbank farming in the case of Mbire), there is need for clear coordination and collaboration of institutions as part of institutional adaptation to shifts in water resource use. This will help institutions to approach water laws and regulations, as well as tackle water-related problems in a more efficient and coordinated manner. Mensah (2014) argues that, from a network governance perspective, coordination and collaboration enhance social and anticipatory learning and action which are vital in building and sustaining institutional resilience. Whilst bringing together a diversity of stakeholders with varying mandates is inherently complex, the advantage in the Mbire context is that institutions already strongly collaborate in flood response and drought relief (led by the RDC). The structural frameworks and foundational processes for coordination towards collaborative water management are therefore present. The existing frameworks and processes of coordination may therefore be adapted to deal with broad water issues along the institutional bricolage concept advanced in the CI approach. This will ultimately facilitate the creation of platforms for collaborative decision-making, bringing together divergent views *vis-à-vis* approach to issues, as well information sharing among institutional players in water management in the area.

6. Conclusions

A clear message from this paper for institutional policy and practice *vis-à-vis* responding to adverse climate dynamics is that institutional adaptation for sustainable water management in arid and semi-arid areas like Mbire can be pursued through institutions embracing complexity thinking and flexibility as well clearly coordinating in all aspects of water management. This will include taking up complexity thinking and clear coordination in demand management (e.g. in issues of water resource conservation and conflict resolution), supply management (e.g. involving issues of infrastructure maintenance and rehabilitation) and planning (e.g. around short- and long-term drought and flood responses). Political and technical support is, however, also critical for successful local institutional adaptation towards sustainable water management. For example, local authorities in water management should be given ample latitude to adapt national regulations to local dynamics. They should also be equipped with technical knowledge and the ability to educate communities on the implications of climate variability and change on water resources and sustainable water use. Improved knowledge of the nature and pace of climate change and its impacts will also aid institutions to take up the right mentality in dealing with climate change issues and identify viable response options.

The research implications of this study are that it clearly shows that there are opportunities for water management and climate adaptation research to jointly contribute knowledge towards improving the capacity of local institutions in managing water resources under increasing climate challenges. Locally derived insights as those coming out of this work hold value for illustrating and understanding better not only the contested nature of water management in the context of increasing climate challenges but also the conditions and factors critical for institutional adaptation towards effective and sustainable water management in specific contexts.

References

Adger, W.N. (2000), "Institutional adaptation to environmental risk under the transition in Vietnam", *Annals of the Association of American Geographers*, Vol. 90 No. 4, pp. 738-758.

Agrawal, A. (2008), *The Role of Local Institutions in Adaptation to Climate Change*, Social Development Department, The World Bank, Washington, DC.

Brown, D., Chanakira, R.R., Chatiza, K., Dhliwayo, M., Dodman, D., Masiiwa, M., Muchadenyika, D., Mugabe, P. and Zvigadza, S. (2012), "Climate change impacts, vulnerability and adaptation in Zimbabwe", IIED Working Paper No 3, International institute for Environment and Development, London, available at: pubs.iied.org/pdfs/10034IIED.pdf

Chagutah, T. (2010), *Climate Change Vulnerability and Adaptation Preparedness in Southern Africa: Zimbabwe Country Report*, Heinrich Boll Stiftung Southern Africa, Cape Town, available at: https://za.boell.org/sites/default/files/downloads/HBF_web_Zim_21_2.pdf

Cleaver, F. (2012), *Development through Bricolage: Rethinking Institutions for Natural Resource Management*, Routledge, London.

Cleaver, F. and de Koning, J. (2015), "Furthering critical institutionalism", *International Journal of the Commons*, Vol. 9 No. 1, pp. 1-18, available at: http://doi.org/10.18352/ijc.605

Davis, R. and Hirji, R. (2014), *Climate Change and Water Resources Planning, Development and Management in Zimbabwe: An Issue Paper*, The World Bank, Washington, DC.

Government of Zimbabwe (1991), *Natural Resources Protection Regulation*, Government of Zimbabwe, Harare.

Government of Zimbabwe (2002), *Environmental Management Act*, Government of Zimbabwe, Harare.

Government of Zimbabwe (2012), *Zimbabwe National Water Policy*, Ministry of Environment, Water and Climate, Harare.

Government of Zimbabwe (2015), *Zimbabwe National Climate Change Response Strategy*, Ministry of Environment, Water and Climate, Harare.

Hall, K., Cleaver, F., Franks, T. and Maganga, F. (2013), "Critical institutionalism: a synthesis and exploration of key themes", Working Paper No 63, Kings College London, Environment, Politics and Development.

Intergovernmental Panel on Climate Change (IPCC) (1996), *Climate Change 1995: Impacts, Adaptations and Mitigations: Contribution of Working Group II to the Second Assessment Report of the IPCC*, Cambridge University Press, Cambridge, MA.

Kujinga, K. (2004), "The dynamics of stakeholder participation in water resources management in Zimbabwe: a case study of the agricultural sector", MSc thesis, University of the Western Cape.

Kwangare, J., Mayo, A. and Hoko, Z. (2014), "Sustainability of donor-funded rural water supply and sanitation projects in Mbire district, Zimbabwe", *Physics and Chemistry of the Earth*, Vols 76/78, pp. 134-139.

Mensah, K.O. (2014), "Linking precaution to adaptive co-management to adapt rural water resources to climate change in Ghana", PhD thesis, University of Guelph, Ontario.

Mtisi, S. (2011), "Water reforms during the crisis and beyond: Understanding policy and political challenges of reforming the water sector in Zimbabwe", ODI Working Paper No 333, Overseas Development Institute, London, available at: www.odi.org/sites/odi.org.uk/files/odi-assets/publications-opinion-files/7241.pdf

Mtisi, S. and Prowse, M. (2012), *Baseline Report on Climate Change and Development in Zimbabwe*, Government of Zimbabwe, Harare.

North, D.C. (1990), *Institutions, Institutional Change and Economic Performance*, Cambridge University Press, Cambridge, MA.

Nyamwanza, A. and Kujinga, K. (2017), "Climate change, sustainable water management and institutional adaptation in rural Sub-Saharan Africa", *Environment, Development and Sustainability*, Vol. 19 No. 2, pp. 693-706.

Sadoff, C. and Muller, M. (2009), "Water management, water security and climate change adaptation: Early impacts and essential responses", Global Water Partnership Technical Committee (TEC), Stockholm, TEC Background Paper No 14.

Simonsen, S.H., Biggs, R., Schluter, M., Schoon, M., Bohensky, E., Cundill, G., Dakos, V., Daw, T., Kotschy, K., Leitch, A., Quinlan, A., Peterson, G. and Moberg, F. (2014), *Applying Resilience Thinking: Seven Principles for Building Resilience in Social-Ecological Systems*, Stockholm Resilience Centre, Stockholm.

Twikirize, D. (2005), "An assessment of traditional water management practices and their implications for improved water governance in the Limpopo Basin: a case of the Sibasa Dam in Mzingwane catchment, Zimbabwe", MSc Thesis, University of Zimbabwe, Harare.

Further reading

Alvera, P. (2013), "The role of indigenous knowledge systems in coping with food security and climate challenges in Mbire district, Zimbabwe", MSc thesis, University of Zimbabwe, Harare.

Mupangwa, J.F., Nyabadza, T., Mberengwa, I. and Chaipa, I. (2006), "Problem animal control strategies in human-elephant conflict areas of Dande Communal Lands, Zimbabwe: a study in sustainability", *Journal of Sustainable Development*, Vol. 8 No. 1, pp. 53-69.

About the author

Admire Mutsa Nyamwanza is a researcher with the Human Sciences Research Council (HSRC), Economic Performance and Development (EPD) Unit, based in Cape Town, South Africa. His main research interests are in the areas of climate change adaptation, livelihoods, resilience and social change in developing countries. Admire Mutsa Nyamwanza can be contacted at: anyamwanza@gmail.com

11

Agro-pastoralists' determinants of adaptation to climate change

Keneilwe Ruth Kgosikoma and Phatsimo Cotildah Lekota
Department of Agricultural Economics, Education and Extension,
Faculty of Agriculture, Botswana University of Agriculture and Natural Resources,
Gaborone, Botswana, and

Olaotswe Ernest Kgosikoma
Department of Agricultural Research,
Ministry of Agricultural Development and Food Security, Gaborone, Botswana

Abstract

Purpose – The purpose of this study is to analyze smallholder farmers' perceptions on climate change and its stressors, their adaptation strategies and factors that influence their adaptation to climate change.

Design/methodology/approach – The study was conducted in Kweneng district, located in the south eastern part of Botswana. Multi-stage sampling was used to obtain a representative sample from three sub-districts in the district. A structured questionnaire was used to collect data by using face-to-face interviews.

Findings – Majority of farmers perceived an increase in mean annual temperature and the number of hot days and a decrease in mean annual rainfall and the number of rainfall days over the past 10 years as indicators of climate change. The prominent adaptation strategies included changes in planting dates for crops and supplementary feeding for livestock. The logistic regression results show that gender, age, household size, poverty, shortage of land, mixed farming and knowledge about climate change significantly influence adaptation.

Practical implications – The findings indicate that climate change policy should target agricultural diversification at the household level and dissemination of information on climate change and adaptation strategies.

Originality/value – Policy recommendations can be suggested: government climate change interventions should target agricultural diversification at the household level, and this study provides insights on what influences adaptation strategies and what should be targeted to build resilience in the agricultural sector.

Keywords Perceptions, Climate change, Resilience, Adaptation, Agricultural diversification, Agro-pastoralists

1. Introduction

The impacts of climate change are prominent worldwide (Oppenheimer *et al.*, 2014), especially in drylands, where its adverse effects are exacerbated by high rainfall variability (Kgosikoma and Batisane, 2014) coupled with high temperatures. Thus, climate change threatens agricultural productivity through increased temperatures, changes in precipitation patterns and increased occurrences of extreme weather conditions (Nelson, 2009), new crop and livestock pests, limited supply of irrigation water and the increased severity of soil erosion (Adams *et al.*, 1998). In addition, climate change may create new and suitable conditions for weeds, insects and pathogens to proliferate, resulting in further decline in agricultural productivity. The competition between weeds and crops for space, water and nutrients from the soil has already been attributed to the highest crop losses globally, about 34 per cent (Oerke, 2006), and may be exacerbated by climate change. Similarly, the productivity of livestock sector is declining because of heat stress, poor nutrition (Muntifering *et al.*, 2006) and shortage of drinking water, which can be attributed to climate change.

Agriculture in Africa supports livelihood of 80 per cent of the population (FAO, 2016), representing over 800 million inhabitants in 2010 (FAO, 2012). Hence, it is a dominant economic activity, particularly to rural households in drylands such as Kgalagadi-Namib region, as it is the main source of food, income and employment. Given the high vulnerability of rural communities in drylands to climate change, it is essential to build resilience to climate change in the agricultural sector through the adoption of climate smart agricultural practices. Sustainable adaptation practices ensure that farmers achieve their food, income and livelihood security objectives in the face of changing climatic and socioeconomic conditions and volatile short-term changes in local and large-scale markets (Phuong, 2011; Kandlinkar and Risbey, 2000), thus reducing vulnerability to climate change (Nhemachena and Hassan, 2007) and poverty (Halsnaes and Traerup, 2009).

The local farming communities have always adapted to perceived environmental risks, and evidence suggests that farmers worldwide acknowledge changes in climatic conditions and its threat to their livelihood. Most farmers in African countries have observed long-term increased temperatures, declining and pattern change in precipitation and increase in drought frequencies changes in rain patterns as a results of climate change (Hassan and Nhemachena, 2008; Gbetibouo, 2009). Farmers' vulnerability and perception to climate change is influenced by factors such as soil fertility, lack of finance, access to water for irrigation and access to climate information (Maddison, 2006). In addition, farmers with high education and farming experience (Gbetibouo, 2009; Hassan and Nhemachena, 2008) and access to extension services and mass-media are likely to have high awareness of climate risks (Sampei and Aoyagi-Usui, 2009) and better adaptive capacity. Farm size, tenure status, access to market and credit availability are other major determinants of adoption in Africa (Maddison, 2006).

Farmers' awareness and perceptions of changes in climatic conditions shape their response to risks associated with climate change. In Botswana, knowledge on farming communities' perception to climate change and determinants of adaptation practices adoption is limited, except in the Okavango region. As a result, this study was conducted to:

- determine farmers' perceptions on climate change and its stressors in Botswana;
- identify farmers' adaptation strategies to climate change; and
- determine factors that influence farmers' adaptation to climate change.

Understanding how Batswana farmers have coped over the years will help policy-makers implement sustainable adaptation strategies that will help reduce climate change impacts in future.

2. Methodology
2.1 Study area
The study was conducted in Kweneng district, located in the south eastern part of Botswana, in 2014. The target population for the study was the Kweneng district

subsistence or smallholder farmers who are highly vulnerable to drought, a key stressor of climate change. Kweneng district is semi-arid with annual rainfall ranging between 300 and 500 mm and mean summer temperature ranging between 24°C and 27°C (Kgosikoma et al., 2012). Kweneng district is generally dominated by non-calcareous sandy soils, and the vegetation type is classified as central bush savanna. Agriculture is the main economic activity in this district, which is essential for local food security and communities' livelihoods.

2.2 Data collection

A structured questionnaire was used to collect primary data from farmers by using face-to-face interviews to make sure farmers understand and are able to respond to the questionnaire, thus maximizing the response rate. This approach is widely used to collect data in ecology and natural resource management (White et al., 2005), including ecological knowledge of resource users. Face-to-face interviews are commonly used when collecting primary data from smallholder farmers because a high response rate is obtained compared to other methods of data collection (Hox and De Leeuw, 1994). In addition, other methods such as telephone or mail survey are not ideal as poor smallholder farmers would not necessarily have access to these.

The questionnaire captured farmers' demographic characteristics, perceptions on climate change including changes in rainfall, temperature and extreme weather events in the past 10 years, and important indicators of climate change. It also captured how climate change has affected crop and livestock production in the past 10 years and the strategies used by farmers to cope with climate change. According to Reyes-García et al., 2015, four main types of local indicators can be derived from local knowledge to explain climate change. These are local observations of climate change (including changes in temperature, precipitation and wind) and its impacts on the physical, biological and socio-economic systems.

Empirical evidence suggests that local people with long history of interaction with their environment develop intricate and complex systems of first-hand knowledge on weather and climate variability, as well as climate change (Orlove et al., 2000; Stigter et al., 2005; Fernández-Llamazares et al., 2015; Marin, 2010). According to Huntington et al., 2004; Rosenzweig and Neofotis, 2013; Fernández-Llamazares et al., 2016, there is an overlap between local knowledge and scientific information, highlighting the critical role of local perceptions in climate change deliberations. Furthermore, farmers are asked about determinants of adaptation strategies.

The study used a multi-stage sampling procedure to obtain a representative sample of the population from three sub-districts of the Kweneng district. The first stage involved listing of villages in the district and then purposively selecting villages dominated mainly by agricultural activity. Within the selected villages, simple random sampling was then used to select a sample of 100 farmers interviewed for this study. The sample size was determined by following the minimum sample size calculation as suggested by Peduzzi et al. (1996). The minimum number of observations included is $N = 10 \, k/p$, where p is the smallest of the proportions of negative or positive observations in the population and k is the number of covariates or independent variables. For this study, $k = 14$ and $p = 0.2$ (proportions of negative observations), and the minimum number of observations (sample size) is 25. However, according to Long (1997), if the resulting number is less than 100, you should increase it to 100 for the logistic regression model.

2.3 Theoretical framework

In the present study, the dependent variable is binary, that is, either the farmer used an adaptation strategy or did not use. A relevant statistical model when the dependent variable is binary is the logistic regression model. Following Uchezumba *et al.* (2009), the choice of binary logistic regression techniques was based on two reasons: first, the technique can be used to analyze the relationship between a categorical response variable and a set of both continuous and categorical variables and second, the technique is best suited for modeling non-linear distribution, which is not appropriate with ordinary least squares. Following Gujarati (2003), a logistic regression model is specified as:

$$P_i = E(Y_i = 1/X_i) = \frac{1}{1 + e^{-\left(\alpha_i + \sum_{i=1}^{K} \beta_i X_i\right)}} \tag{1}$$

where P_i is the probability of household i adopting at least one adaptation strategy, Y_i is the level of adaptation by the same household i, Xi is a set of explanatory variables influencing the participation of household i in the cattle market and the β_i,s are the parameters to be estimated.

The term $\left(\alpha_i + \sum_{i=1}^{K} \beta_i X_i\right)$ can be denoted as Z_i, so that equation (1) becomes:

$$P_i = \frac{1}{1 + e^{-Z_i}} \tag{2}$$

Given that the probability of adopting at least one adaptation strategy (P_i) is as given in equation (2), then the probability of not adapting any strategy $(1 - P_i)$ can be expressed as specified below:

$$\frac{1}{(1 - P_i)} = 1 + e^{Z_i} \tag{3}$$

The odds ratio $P_i/1 - P_i$ is, therefore, is given as:

$$\frac{P_i}{1 - P_i} = \frac{1 + e^{Z_i}}{1 + e^{-Z_i}} \tag{4}$$

Taking the logarithm of equation (4), the logit model takes the form:

$$L_i = \ln\left[\frac{P_i}{1 - P_i}\right] = P_0 + \sum_{i=1}^{k} \gamma_i X_i + \varepsilon_i \tag{5}$$

where L_i is the logit and ε_i is the error term, and the other variables are defined as before. The marginal effects for the binary variables is calculated by predicting the outcome probability for each observation given that adaptation = 1 and then again for each observation substituting adaptation = 0. The sample average of the difference between those outcome probabilities is the average marginal effect or just marginal effect. The marginal effect for the categorical variables on the probability of household i adapting to climate change is determined by taking the partial derivative of the probability of the outcome with respect to explanatory variable for each observation in the data set. The sample average of that is then reported as the average marginal effect.

The binominal logit model was used to determine the factors that influence farmers to adapt to climate change. The diagnostic tests of the model showed high correlation of 0.80 between some covariates in the initial model, resulting in some variables being dropped. The model was tested for common regression model problems, that is, model specification, model fit and multicollinearity, and there were no indications of any of these problems. The variance inflation factor (VIF) often used to identify multicollinearity indicated that the largest VIF was only 2.86. A VIF of 10 indicates presence of multicollinearity, which requires attention. The probability model was correctly specified and fit the data well according to the Hosmer and Lemeshow's goodness-of-fit test and the STATA "linktest" diagnostic tests, which produced statistically insignificant results (probability $> \chi^2 =$ 0.9707; $p > |z| = 0.329$), indicating that the model fits the data well and has no specification error. A model with Huber–White robust standard errors was adopted to counter any heteroscedasticity problems.

2.4 Empirical framework
To evaluate the determinants of adaptation to climate change, the following general logistic regression model was used:

$$Logit\,(P_i) = \ln(P_i/1 - P_i) = \beta_0 + \beta_i X_i + \ldots + \beta_n X_n \qquad (6)$$

where $\ln(P_i/1 - P_i)$ is the logit for adaptation to climate change choices; P_i represents adaptation; $1 - P_i$ is not adapting and X_is represents covariates as previously stated. The empirical model, with the explanatory variables selected based on theory, is presented as:

$$\ln(P_i/1 - P_i) = \beta_0 + \beta_1 FD_i + \beta_2 FE_i + \beta_3 FP_i + \varepsilon_i \qquad (7)$$

The explanatory variables hypothesized to influence farmers' ability to adapt can be broadly categorized into demographic characteristics (*FD*), endowment and (*FE*) and perceptions on adaptation constraints (*FP*) and are described in Table I and subsequently discussed.

3. Results
3.1 Farmers' perceptions on climate change and its stressors
The farming community in Kweneng had observed several indictors and impacts associated with climate change (Table II). The majority of the respondents in the study indicated that the temperature and the number of hot days have increased over the past 10 years by 97 and 91 per cent, respectively. Almost all farmers in Kweneng have also observed decline in rainfall, and 95 per cent of them have noticed a decrease in rainfall days. Most agro-pastoralists in Kweneng were concerned with reoccurrence of drought, particularly that their observed trends indicated increased drought frequency. Based on most farmers' perceptions, flood occurrence has not changed much in the past 10 years. The observed changes in climatic conditions reported by farmers were associated with reduced crop and livestock productivity.

Farmers attributed decreased crop and livestock productivity to several stressors associated with climate change. Most farmers identified drought and low rainfall as the major risks to agricultural productivity (Figure 1). In addition, high temperature was

Table I. Description of explanatory variables hypothesized to influence adaptation

Variable	Full variable name	Description	Hypothesized sign
GENDER	Gender of a household head	Binary, 1 if men and 0 otherwise	+
AGE	Age of a household head	Categorical	−
HHSize	Household size	Continuous	+
YrOFFARMIN	Years of farming	Continuous	+
EDUC	Years of education for a household head	Continuous	+
AGRICINC	Income from agriculture	Continuous	+
NON_AGRIC_INC	Income from other sources	Binary, 1 if there are other sources of income, 0 otherwise	+
MIXEDFARM	Farmer practices mixed farming	Binary, 1 if mixed farming is practiced, 0 otherwise	+
KNOWCLIMATE	Farmer knows about climate change	Binary, 1 if farmer knows about climate change, 0 otherwise	+
LACK_KNWLGD	Lack of knowledge on adaptation strategies an important constraint to adaptation	Binary, 1 if important, 0 otherwise	−
LACK_CREDIT	Lack of access to credit an important constraint to adaptation	Binary, 1 if important, 0 otherwise	−
POVERTY	Poverty an important constraint to adaptation	Binary, 1 if important, 0 otherwise	−
LACK_WATER	Lack of access to water an important constraint to adaptation	Binary, 1 if important, 0 otherwise	−
LACK_LAND	Lack or shortage of agricultural land is an important constraint to adaptation	Binary, 1 if important, 0 otherwise	−

Table II. Farmers' perceptions of climate change and its impacts on production as a percentage of the total respondents ($N = 91$)

Perceived climatic changes and its impacts on agriculture during the past decade	Percentage			
	Increase	No change	Decrease	Unsure
Mean temperature	97	0	1	2
Number of hot days	91	5	1	2
Mean rainfall	2	1	97	0
Number of rainfall days	3	1	95	1
Occurrence of drought	82	11	1	5
Occurrence of floods	1	74	14	11
Crop productivity	14	3	76	0
Livestock productivity	18	1	73	3

reported to cause poor growth of crops and livestock because of heat stress. A moderate proportion of farmers mentioned pests and diseases as climate change stressors that result in reduced agricultural productivity. Poor vegetation was mentioned by only a negligible proportion of Kweneng farmers as a climate change stressor that leads to decline in crop and livestock productivity.

3.2 Adaptation strategies used by farmers

From a sample of 91 farmers, the majority of the farmers (82 per cent) had adapted to climate change. The adaptation strategy used by the majority of the crop farmers in Kweneng district was to change of planting dates to be aligned with the current rainfall patterns (months). Other crop-related adaptation strategies included change in crop varieties planted, switching from crop to livestock production, implementation of soil conservation techniques, use of irrigation and use of shades and shelters and changes in the use of chemical fertilizers, pesticides or insecticides. Livestock production adaptation strategies farmers perceived as appropriate in the region were vaccinating farm animals, supplementary feeding, fencing and shading (housing) (Table III).

3.3 Determinants of adaptation to climate change

From the results of the logistic regression model, the determinants of adaptation to climate change are gender, age, household size, mixed farming, knowledge about climate change, poverty and shortage of land. The results indicated that female-headed households are 16 per cent more likely to adapt to climate change than male-headed households. Moreover, the results revealed that increased households' size and age of head negatively influenced farmer's adaptive capacity. Mixed farming and knowledge on climate change increased farmer's adaptive capacity by 18.3 and 26 per cent, respectively. Other significant determinants of adaptation to climate change identified by farmers in Kweneng were

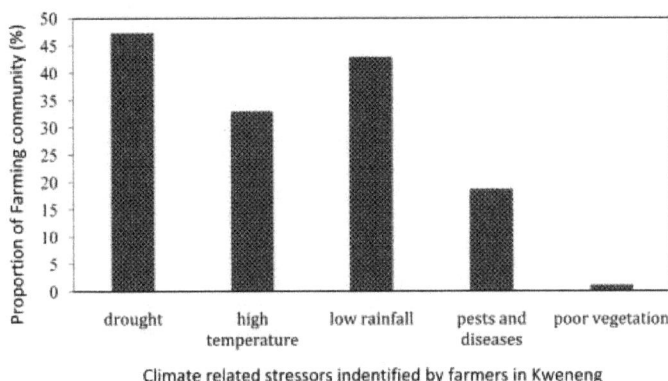

Figure 1. Climate change-related stressors that reduce agricultural productivity as perceived by farmers

Table III. Adaptation strategies used by farmers

Adaptive strategy	Frequency	(%)
Crop sector		
Change of planting dates	64	85
Change crop variety	12	23
Switching from crops to livestock	1	1
Implementation of soil conservation techniques	20	27
Use of irrigation	3	4
Change use of chemical fertilizers, pesticides or insecticides	15	20
Livestock sector		
Supplementary feeding	5	5
Vaccinations	1	1
Use of shades and shelters	3	4

poverty and shortage of land, which individually reduced adaptive capacity by about 20 per cent (Table IV).

4. Discussion
4.1 Farmers' perceptions on climate change and its stressors
Majority of agro-pastoralists in Kweneng associated observed increased temperature with climate change. This view is augmented by increased number of hot days experienced in that area and consistent with projections that temperatures in semi-arid of southern Africa will increase by between 3.4°C and 4.2°C, which is more than the 1981-2000 average under the A2 scenario by end of the twenty-first century (Niang et al., 2014). In addition, farmers suggested that annual rainfall and number of rainy days have decreased because of climate change. Subsequently, farmers mentioned that drought frequency has increased, and their observations are supported by other studies in the region (Makhado et al., 2014). High proportions of farming community in Kweneng associated observed climate changes with decreased agricultural productivity, including both crop and livestock sectors. This could be explained by the fact that agricultural production systems in Botswana and southern Africa are largely dependent on rainfall (Makhado et al., 2014) and thus vulnerable to rainfall variability, as suggested in other studies (Kolawole et al., 2016). Similarly, it has been demonstrated that rainfall variability drives both crop yields (Kolawole et al., 2016) and livestock productivity (Kgosikoma and Batisane, 2014) elsewhere in Botswana. The livelihood of smallholder livestock farmers in communal lands of Botswana is therefore more vulnerable to climate change, partly because of compounding effect of land degradation and partly because of insecure land tenure (Dougill et al., 2010).

As suggested by farmers in Kweneng, drought and low rainfall are the primary climate-related stressors to agricultural sector. Similarly, it was reported that drought and low rainfall have high negative impact on crop failure, especially maize and sorghum in the Okavango region of Botswana (Kolawole et al., 2016). Frequent drought also causes decline in livestock body condition and eventually increased mortality, as observed in other drylands

Table IV. Determinants of smallholder farmers' adaptation to climate change

Variable	Coefficient (robust standard error)	Marginal effect
GENDER	−1.846 (1.088)*	−0.162*
AGE	−1.180 (0.654)*	−0.104*
HHSize	−0.251 (0.105)**	−0.022**
YrOFFARMIN	0.016 (0.024)	0.001
EDUC	0.064 (0.154)	0.006
AGRICINC	0.0004 (0.0003)	0.00004
NON_AGRIC_INC	1.318 (1.084)	0.116
MIXEDFARM	2.082 (1.015)**	0.183**
KNOWCLIMATE	2.957 (1.421)**	0.260
LACK_KNWLGD	1.359 (1.091)	0.119
CREDIT	1.716 (1.608)	0.151
POVERTY	−2.304 (1.022)**	−0.203**
WATER	1.838 (1.344)	0.162
LAND	−2.264 (0.931)**	−0.199***
N	91	
Wald χ^2 (15)	25.19	
Probability $> \chi^2$	0.05	

Notes: ***; ** and * indicate significance at 1%, 5% and 10% probability levels, respectively

(Opiyo *et al.*, 2015. In addition, high temperatures are also associated with low output from agriculture sector because of high water demand and heat stress. In the livestock sector, exotic higher-producing breeds that are suited to farming in temperate climate are more vulnerable to heat stress, and their productivity is likely to decline more than indigenous breeds (Archer, 2011). Simulations demonstrated that increased temperature by 2°C and 3°C leads to reduced maize yields by 21 and 36 per cent and sorghum yields by 16 and 26 per cent, respectively, in Botswana (Chipanshi *et al.*, 2004), which is consistent with farmers' perception.

4.2 Adaptation strategies used by farmers

The food security and livelihood of agro-pastoral communities are threatened by climate change, and innovative interventions are necessary to improve agricultural resilience. Kweneng farmers reported using a variety of adaptation strategies to minimize the risks of observed climate change in their production, just like other farmers in Botswana (Kgosikoma and Batisane, 2014; Mogotsi *et al.*, 2011) and other drylands (Opiyo *et al.*, 2015). They indicated that planting dates had been adjusted in response to late rain onset and further enhanced by change in crop variety. Drought-tolerant and early-maturing crop varieties are highly recommended for drylands and have been applied by other farmers in southern Africa (Wiid and Ziervogel, 2012). Investment and research innovation are needed to develop new crop varieties, including hybrids that are highly tolerant to temperature, moisture stress and other relevant climatic conditions (Smit and Skinner, 2002). Some farmers also suggested the use of soil conservation technique to protect the soil from degradation and maintain its productivity as an adaptive strategy.

Overall, smallholder livestock adaptive capacity among Kweneng farmers was low as only few practices were suggested, and therefore, more needs to be done to build adaptive capacity in this sector. Supplementation was suggested by few farmers, and that could be because government subsidizes livestock feed during drought periods. In addition, indigenous browser plants have high potential as feed (Aganga *et al.*, 2000) to be used to supplement livestock.

4.3 Determinants of adaptation to climate change

The logistic regression model results highlighted several factors as determinants of adaptation to climate, including gender of the household head. Contrary to expectation, female-headed households are more likely to adapt to climate change than male-headed households, and this could partly be attributed to willingness of women to change their livelihood strategy in an effort to support their families. In addition, age of household head negatively affected the adaptation to climate change. A plausible explanation is that older farmers may be more conservative and more risk-averse compared to younger farmers, resulting in a lower likelihood of adopting new technologies (Gbetibouo, 2009). The results also indicated that large family size also increased farmers' vulnerability to climate change as a unit increase in the household size resulted in a 2.2 per cent reduction in the probability of adapting to climate change. That is because a large family has high consumption demand, and this put enormous pressure on little resources available during drought periods, and some families may be forced to divert part of the labor force to off-farm activities in an attempt to earn income.

The results showed that knowledge about climate change increases the probability of adaptation by 26 per cent. Similar findings were reported by Atinkut and Mebrat (2016), who found a positive and significant relationship between access to information on climate change and adaptation. As a result, improved extension services that provide technical support on agriculture and climate change services will significantly reduce vulnerability to climate risk (Harvey *et al.*, 2014). Farmers need to be educated on the vulnerability of specific species/crops and the appropriate species/crop mix, including drought-resistant breeds/ crops so that they can adopt appropriate adaptation practices to minimize the adverse impact of climate change (Kabubo-Mariara, 2008).

Diversification of herd composition or crops is an essential component of adaptation to climate risk in the agricultural sector (Smit and Skinner, 2002) because of improved access to market and basic food (Opiyo *et al.*, 2015). The results of this study also confirmed that agro-pastoralists with diversified agricultural practices (mixed farming) are more (18.3 per cent) resilient to climate shocks than those who practice either crop or livestock production only. Unfortunately, smallholder farmers normally have limited resources to enhance diversification and as a result are more vulnerable to climate risks (Harvey *et al.*, 2014). Poverty was therefore identified by farmers as a strong determinant of adaptation to climate change. Limited access to resources such as land also contributed significantly toward low adaptive capacity of smallholder farmers.

5. Conclusions

This study has shown that most farmers in Kweneng district are aware of the increasing temperatures and decrease in rainfall and have attempted to adapt different strategies to mitigate the effects of the changing climate. The predominant adaptation strategies used by crop farmers were changes in planting dates in line with shifts in rainfall season onset, changes in crop varieties planted, changes in the use of chemical fertilizer, pesticides and insecticides, implementation of soil conservation techniques and irrigation. The adaptation strategies pointed out as appropriate for use by livestock farmers were supplementary feeding, vaccination and provision of shading or livestock housing. However, smallholder livestock adaptive capacity among Kweneng farmers was low as only few practices were suggested, and therefore, there is a need to build adaptive capacity in this sector.

The binary logit model results indicated that gender, age, household size, poverty and lack of access to credit significantly and negatively affect adaptation to climate change, whereas diversified agricultural practices and knowledge of climate change significantly and positively influence adaptation. Given the significance of knowledge about climate change on adaptation, government should implement programs that will help increase access to information on climate change and the appropriate adaptive strategies. Policy options to facilitate the availability of credit; investment on yield-increasing technologies; opportunities for off-farm employment; research on the use of new crop varieties and livestock breeds that are more suited to drier conditions; and investment in irrigation should be implemented to help increase production and decrease the vulnerability of farmers toward climate change.

References

Adams, R.M., McCarl, B.A., Segerson, K., Rosenzweig, C., Bryant, K.J., Dixon, B.L., Conner, R., Evenson, R.E. and Ojima, D. (1998), "The economic effects of climate change on US agriculture", in Mendelsohn, R. and Neumann, J. (Eds), Chapter 2, *The Economics of Climate Change*, Cambridge University Press, Cambridge, MA.

Aganga, A.A., Omphile, U.J., Chabo, R.G. and Kgosimore, M. (2000), "Available feed resources to goats under communal management in Gaborone-Agriculture region in Botswana", in Aganga, A.A., Kgosimore, M., Omphile, U.J. and Chabo, R.G. (Eds), *Optimal Feeding of Grazing Animals in Botswana. Proceeding of the Livestock Feeding Systems Workshop held at the Center for Continuing Education*, Botswana College of Agriculture, Gaborone, 28-31 May.

Atinkut, B. and Mebrat, A. (2016), "Determinants of farmers choice of adaptation to climate variability in Dera woreda, South Gondar zone, Ethiopia", *Environmental Systems Research*, Vol. 5 No. 1, doi: 10.1186/s40068-015-0046-x.

Chipanshi, A.C., Chanda, R. and Totolo, O. (2004), "Vulnerability assessment of the maize and sorghum crops to climate change in Botswana", *Climatic Change*, Vol. 61, pp. 339-360.

Dougill, A.J., Fraser, E.D.G. and Reed, M.S. (2010), "Anticipating vulnerability to climate change indryland pastoral systems: using dynamic systems models for the Kalahari", *Ecology and Society*, Vol. 15 No. 2, p. 17, available at: www.ecologyandsociety.org/vol15/iss2/art17/

FAO (2012), *FAO Statistical Yearbook 2012: Africa Food and Agriculture*, FAO, Rome.

FAO (2016), "Diversification strategies and adaptation deficit: evidence from rural communities in Niger", in Asfaw, S., Palma, A. and Lipp er, L. (Eds), ESA Working Paper No. 16-02, FAO, Rome.

Fernández-Llamazares, Á., Díaz-Reviriego, I., Guèze, M., Cabeza, M., Pyhälä, A. and Reyes-García, V. (2016), "Local perceptions as a guide for the sustainable management of natural resources: empirical evidence from a small-scale society in Bolivian Amazonia", *Ecology and Society*, Vol. 21 No. 1, p. 2, available at: http://dx.doi.org/10.5751/ES-08092-210102

Fernández-Llamazares, Á., Méndez-López, E., Díaz-Reviriego, I., McBride, M., Pyhälä, A., Rosell-Melé, A. and Reyes-García, V. (2015), "Links between media communication and local perceptions of climate change in an indigenous society", *Climate Change*, Vol. 131 No. 2, pp. 307-320.

Gbetibouo, G.A. (2009), "Understanding Farmers' Perceptions and Adaptations to Climate Change and Variability: the Case of the Limpopo Basin, South Africa", IFPRI Discussion Paper No. 00849.

Gujarati, D.N. (2003), *Basic Econometrics*, McGraw Hill.

Halsnaes, K. and Traerup, S. (2009), "Development and climate change: a mainstreaming approach for assessing economic, social, and environmental impacts of adaptation measures", *Environmental Management*, Vol. 43 No. 5, pp. 765-778, doi: 10.1007/s00267-009-9273-0.

Harvey, C.A., Rakotobe, Z.L., Rao, N.S., Dave, R., Razafimahatratra, H., Rabarijohn, R.H., Rajaofara, H. and MacKinnon, J.L. (2014), "Extreme vulnerability of smallholder farmers to agricultural risks and climate change in Madagascar", *Philosophical Transactions of the Royal Society of London. Series B, Biological Sciences*, Vol. 369 No. 1639, p. 20130089, doi: 10.1098/rstb.2013.0089.

Hassan, R. and Nhemachena, C. (2008), "Determinants of African farmers' strategies for adapting to climate change: multinomial choice analysis", *African Journal of Agricultural and Resource Economics*, Vol. 2 No. 1, pp. 83-104.

Hox, J.J. and De Leeuw, E.D. (1994), *Quality and Quantity*, Vol. 28 No. 4, pp. 329-344, available at: https://doi.org/10.1007/BF01097014

Huntington, H.P., Callaghan, T., Fox, S. and Krupnik, I. (2004), "Matching traditional and scientific observations todetect environmental change: a discussion on Arctic terrestrial ecosystems", *AMBIO*, Vol. 13, pp. 18-23.

Kabubo-Mariara, J. (2008), "Climate change adaptation and livestock activity choices in Kenya: an economic analysis", *Natural Resources Forum*, Vol. 32 No. 2, pp. 131-141.

Kandlinkar, M. and Risbey, J. (2000), "Agricultural impacts of climate change: if adaptation is the answer, what is the question?", *Climatic Change*, Vol. 45, pp. 529-539.

Kgosikoma, O.E. and Batisane, N. (2014), "Livestock population dynamics and pastoral communities' adaptation to rainfall variability in communal lands of Kgalagadi South, Botswana", *Pastoralism: Research, Policy and Practice*, Vol. 4 No. 1, p. 19, doi: 10.1186/s13570-014-0019-0.

Kgosikoma, O.E., Harvie, B.A. and Mojeremane, W. (2012), "Bush encroachment in relation to rangeland management systems and environmental conditions in Kalahari ecosystem of Botswana", *African Journal of Agricultural Research*, Vol. 17, pp. 2312-2319.

Kolawole, O.D., Motsholapheko, M.R., Ngwenya, B.N. and Thakadu, O. (2016), "Climate variability and rural livelihoods: how households perceive and adapt to climatic shocks in the Okavango Delta, Botswana", *Weather, Climate, and Society*, Vol. 8, pp. 131-145.

Long, J.S. (1997), *Regression Models for Categorical and Limited Dependent Variables*, Sage Publications, Thousand Oaks, CA.

Maddison, D. (2006), "The perception of and adaptation to climate change in Africa", CEEPA Discussion Paper No. 10. Centre for Environmental Economics and Policy in Africa, University of Pretoria.

Makhado, R.A., Saidi, A.T. and Tshikhudo, P.P. (2014), "Optimising drought adaptation by smallscale farmers in Southern Africa through integration of indigenous and technologically-driven practices", *African Journal of Science, Technology, Innovation and Development*, Vol. 6 No. 4, pp. 265-273.

Marin, A. (2010), "Riders under storms: contributions of nomadic herders' observations to analysing climate change in Mongolia", *Global Environmental Change*, Vol. 20 No. 1, pp. 162-176.

Mogotsi, K., Nyangito, M.M. and Nyariki, D.M. (2011), "The perfect drought? Constraints limiting Kalahari agro-pastoral communities from coping and adapting", *African Journal of Environmental Science and Technology*, Vol. 5 No. 3, pp. 168-177.

Muntifering, R.B., Chappelka, A.H., Lin, J.C., Karnosky, D.F. and Somers, G.L. (2006), "Chemical composition and digestibility of trifolium exposed to elevated ozone and carbon dioxide in a free air (FACE) fumigation system", *Functional Ecology*, Vol. 20 No. 2, pp. 269-275.

Nelson, G.C. (Ed.) (2009), "Agriculture and climate change: an agenda for negotiation in Copenhagen", 2020 Focus No. 16. May 2009, available at: www.ifpri.org/2020/focus/focus16.asp (accessed 5 September 2014).

Nhemachena, C. and Hassan, R. (2007), "Micro-level analysis of farmers' adaptation to climate change in Southern Africa", IFPRI Discussion Paper 00714, International Food Policy Research Institute, Washington, DC.

Niang, I., Ruppel, O.C., Abdrabo, M.A., Essel, A., Lennard, C., Padgham, J. and Urquhart, P. (2014), *Africa*, In Climate Change 2014: Impacts, Adaptation, and Vulnerability. Part B: Regional Aspects. Contribution of Working Group II to the Fifth Assessment Report of the Intergovernmental Panel on Climate Change [Barros, V.R., C.B. Field, D.J. Dokken, M.D. Mastrandrea, K.J. Mach, T.E. Bilir, M. Chatterjee, K.L. Ebi, Y.O. Estrada, R.C. Genova, B. Girma, E.S. Kissel, A.N. Levy, S. MacCracken, P.R. Mastrandrea, and L.L. White (Eds)], Cambridge University Press, Cambridge, United Kingdom and New York, NY, USA, pp. 1199-1265.

Oerke, E.C. (2006), "Crop losses to pests", *The Journal of Agricultural Science*, Vol. 144 No. 1, pp. 31-43.

Opiyo, F., Wasonga, O., Nyangito, M., Schilling, J. and Munang, R. (2015), "Drought adaptation and coping strategies among the turkana pastoralists of Northern Kenya", *International Journal of Disaster Risk Science*, Vol. 6 No. 3, pp. 295-309, doi: 10.1007/s13753-015-0063-4.

Oppenheimer, M., Campos, M., Warren, R., Birkmann, J., Luber, G., O'Neill, B. and Takahashi, K., (2014), "Emergent risks and key vulnerabilities", in Climate Change 2014: Impacts, Adaptation, and Vulnerability. Part A: Global and Sectoral Aspects. Contribution of Working Group II to the Fifth Assessment Report of the Intergovernmental Panel on Climate Change [Field, C.B., V.R. Barros, D.J. Dokken, K.J. Mach, M.D. Mastrandrea, T.E. Bilir, M. Chatterjee, K.L. Ebi, Y.O. Estrada, R.C. Genova, B. Girma, E.S. Kissel, A.N. Levy, S. MacCracken, P.R. Mastrandrea, and L.L. White (eds)]. Cambridge University Press, Cambridge, United Kingdom and New York, NY, USA, pp. 1039-1099.

Orlove, B., Chiang, J. and Cane, M. (2000), "Forecasting Andean rain-fall and crop yield from the infuence of El Niño onPleiades visibility", *Nature*, Vol. 403, pp. 68-71.

Peduzzi, P., Concato, J., Kemper, E., Holford, T.R. and Feinstein, A.R. (1996), "A simulation study of the number of events per variable in logistic regression analysis", *Journal of Clinical Epidemiology*, Vol. 49 No. 12, pp. 1373-1379.

Phuong, L. (2011), "Climate change and farmers' adaptation: a case study of mixed-farming systems in the coastal area in Trieu Van commune, Trieu Phong district, Quang Tri province, Vietnam", Masters thesis, Swedish University of Agricultural Sciences.

Reyes-García, V., Fernández-Llamazares, A., Guèze, M., Garcés, A., Mallo, M., Vila-Gómez, M. and Vilaseca, M. (2015), "Local indicators of climate change: the potential contribution of local knowledge to climate research", *Wiley Interdisciplinary Reviews: Climate Change*, Vol. 7 No. 1, pp. 109-124, doi: 10.1002/wcc.374.

Rosenzweig, C. and Neofotis, P. (2013), "Detection and attribution of anthropogenic climate change impacts", *Wiley Interdisciplinary Reviews: Climate Change*, Vol. 4 No. 2, pp. 121-150, doi: 10.1002/wcc.209.

Sampei, Y. and Aoyagi-Usui, M. (2009), "Mass-media coverage, its influence on public awareness of climate change issues, and implications for Japan's national campaign to reduce greenhouse gas emissions", *Global Environmental Change*, Vol. 19 No. 2, pp. 203-212, doi: 10.1016/j.gloenvcha.2008.10.005.

Smit, B. and Skinner, M. (2002), "Adaptation options in agriculture to climate change: a typology", *Mitigation and Adaptation Strategies for Global Change*, Vol. 7, pp. 85-114.

Stigter, C.J., Zheng, D.W., Onyewotu, L.O.Z. and Mei, X.R. (2005), "Using traditional methods and indigenous technologies for coping with climate variability", *Climate Change*, Vol. 70 Nos 1/2, pp. 255-271.

White, P.C.L., Jennings, N.V., Renwick, A.R. and Baker, N.H.L. (2005), "Questionnaires in ecology: a review of past use and recommendations for best practice", *Journal of Applied Ecology*, Vol. 42, pp. 421-430.

Wiid, N. and Ziervogel, G. (2012), "Adapting to climate change in South Africa: commercial farmers' perception of and response to changing climate", *South African Geographical Journal*, Vol. 94 No. 2, pp. 152-173.

Further reading

Dixon, J., Tanyeri-Abur, Y. and Wattenbach, H. (2004), "Smallholders, globalization and policy analysis", Agricultural Management, Marketing and Finance service occasional paper 5, *Agricultural Support Systems Division*, FAO, Rome.

Fosu-Mensah, B.Y., Vlek, P.L.G. and Manschadi, A.M. (2010), *Farmers' Perception and Adaptation to Climate Change; a Case Study of Sekyedumase District in Ghana*, Center for Development Research (ZEF), University of Bonn, Bonn.

Fussel, H.M. (2007), "Adaptation planning for climate change: concepts, assessment approaches, and key lessons", *Sustainability Science*, Vol. 2 No. 2, pp. 265-275, doi: 10.1007/s11625-007-0032-y.

Hausman, J. and McFadden, D. (1984), "Specification tests for the multinomial logit model", *Econometrica*, Vol. 52 No. 5, pp. 1219-1240.

IPCC (2013), "Summary for policymakers. in: climate change 2013: the physical science basis", Contribution of Working Group I to the Fifth Assessment Report of the Intergovernmental Panel on Climate Change [Stocker, T.F., D. Qin, G.-K. Plattner, M. Tignor, S.K. Allen, J. Boschung, A. Nauels, Y. Xia, V. Bex and P.M. Midgley (eds)], Cambridge University Press, Cambridge, United Kingdom and New York, NY, USA.

Sultan, B. (2012), *Environmental Resource Letter 7: PERSPECTIVE: Global Warming Threatens Agricultural Productivity in Africa and South Asia*, IOP Publishing Ltd Printed in the UK, Universiẗe Pierre et Marie Curie, Paris, France IRD – LOCEAN/IPSL.

Tizale, C.Y. (2007), "The dynamics of soil degradation and incentives for optimal management in the central highlands of Ethiopia", PhD thesis, Department of Agricultural Economics, Extension and Rural Development, Faculty of Natural and Agricultural Sciences, University of Pretoria.

World Bank (2010), *Botswana Climate Variability and Change: Understanding the Risks*, Draft Policy Note.

Corresponding author

Keneilwe Ruth Kgosikoma can be contacted at: kkgosikoma@gmail.com

PERMISSIONS

LIST OF CONTRIBUTORS

Jeetendra Prakash Aryal
International Maize and Wheat Improvement Center (CIMMYT), El Batan, Mexico

M. L. Jat and Tek B. Sapkota
International Maize and Wheat Improvement Center (CIMMYT), New Delhi, India

Arun Khatri-Chhetri
CGIAR Research Program on Climate Change Agriculture and Food Security (CCAFS), Borlaug Institute for South Asia (BISA), New Delhi, India

Menale Kassie
International Centre of Insect Physiology and Ecology (ICIPE), Nairobi, Kenya

Dil Bahadur Rahut
Socioeconomics Program, International Maize and Wheat Improvement Center (CIMMYT), El Batan, Mexico

Sofina Maharjan
International Maize and Wheat Improvement Center (CIMMYT), New Delhi, India

Iordanis Eleftheriadis and Evgenia Anagnostopoulou
Department of Business Administration, University of Macedonia, Thessaloniki, Greece

Rory A. Walshe
Department of Geography, King's College London, London, UK
Institute of Risk and Disaster Reduction (IRDR), University College London, London, UK

Denis Chang Seng
Intergovernmental Oceanographic Commission, UNESCO, Paris, France

Adam Bumpus
School of Geography, Faculty of Science, University of Melbourne, Melbourne, Australia

Joelle Auffray
Apidae Development Innovations Pty Ltd., Melbourne, Australia

Qi Gao
KoGuan Law School, Shanghai Jiao Tong University, Shanghai, China

Md. Zakir Hossain and Md. Ashiq Ur Rahman
Urban and Rural Planning Discipline, Khulna University, Khulna, Bangladesh

Chandan Kumar Jha, Vijaya Gupta, Utpal Chattopadhyay and Binilkumar Amarayil Sreeraman
National Institute of Industrial Engineering, Mumbai, India

Phan N. Duy
Department of Urban Planning, University of Architecture of Ho Chi Minh City, Ho Chi Minh City, Vietnam and School of Geography, Earth and Environment Sciences, University of Birmingham, Birmingham, UK

Lee Chapman
School of Geography, Earth and Environment Sciences, University of Birmingham, Birmingham, UK

Miles Tight
School of Civil Engineering, University of Birmingham, Birmingham, UK

Phan N. Linh
Department of Urban Planning, University of Architecture of Ho Chi Minh City, Ho Chi Minh City, Vietnam

Le V. Thuong
Southern Institute for Spatial Planning of Vietnam, Ho Chi Minh City, Vietnam

Tijana Crnčević
Institute of Architecture and Urban and Spatial Planning of Serbia, Belgrade, Serbia

Violeta Orlović Lovren
Faculty of Philosophy, University of Belgrade, Belgrade, Serbia

Yimer Mohammed
Hawassa University, Hawassa, Ethiopia and Dilla University, Dilla, Ethiopia

Fantaw Yimer and Menfese Tadesse
Hawassa University, Hawassa, Ethiopia

Kindie Tesfaye
International Maize and Wheat Improvement Center (CIMMYT), Addis Ababa, Ethiopia

Admire Mutsa Nyamwanza
Economic Performance and Development Unit, Human Sciences Research Council, Cape Town, South Africa

Keneilwe Ruth Kgosikoma and Phatsimo Cotildah Lekota
Department of Agricultural Economics, Education and Extension, Faculty of Agriculture, Botswana University of Agriculture and Natural Resources, Gaborone, Botswana

Olaotswe Ernest Kgosikoma
Department of Agricultural Research, Ministry of Agricultural Development and Food Security, Gaborone, Botswana

Index

www.ingramcontent.com/pod-product-compliance
Lightning Source LLC
Chambersburg PA
CBHW062000190326
41458CB00009B/2927